河海大学重点立项教材

本书获得河海大学“学院高质量内涵式发展专项经费”（项目编号:424626）资助

跨文化沟通视域下的水外交典型案例

祝吉芳　祝加贝　谷俊仪　编

河海大学出版社
HOHAI UNIVERSITY PRESS
·南京·

图书在版编目(CIP)数据

跨文化沟通视域下的水外交典型案例 = Case Studies of Water Diplomacy: An Intercultural Communication Perspective : 英文 / 祝吉芳, 祝加贝, 谷俊仪编. -- 南京 : 河海大学出版社, 2025. 7.
ISBN 978-7-5630-9729-6

Ⅰ. TV213.4

中国国家版本馆 CIP 数据核字第 2025C1J039 号

书　　名	跨文化沟通视域下的水外交典型案例 KUAWENHUA GOUTONG SHIYU XIA DE SHUIWAIJIAO DIANXING ANLI
书　　号	ISBN 978-7-5630-9729-6
责任编辑	周　贤
特约校对	吕才娟
封面设计	徐娟娟
出版发行	河海大学出版社
地　　址	南京市西康路 1 号(邮编:210098)
电　　话	(025)83737852(总编室)　(025)83722833(营销部)
经　　销	江苏省新华发行集团有限公司
排　　版	南京布克文化发展有限公司
印　　刷	广东虎彩云印刷有限公司
开　　本	718 毫米×1000 毫米　1/16
印　　张	14
字　　数	332 千字
版　　次	2025 年 7 月第 1 版
印　　次	2025 年 7 月第 1 次印刷
定　　价	68.00 元

前言

Preface

“我住江之头，君住江之尾。彼此情无限，共饮一江水。……彼此地相连，依山复靠水。……山山皆北向，条条南流水。”新中国第二任外交部部长陈毅仿宋代词人李之仪的词而作的《赠缅甸友人》，不仅表达了中缅两国人民之间的深厚友谊，还充分彰显出元帅杰出的外交才华。诗中的“江”，为中国境内的澜沧江，澜沧江的下游是湄公河，澜沧江—湄公河绵延 4 880 公里，从中国发源，流经缅甸、老挝、泰国、柬埔寨、越南，形成了“一江连六国”的自然胜景；沿岸民众“共饮一江水”，世代依水繁衍生息，孕育出各具特色又相亲相近的文化。

然而，进入 21 世纪，在经济发展、人口增长、环境污染、气候变化等多重因素的叠加影响下，湄公河流域成为全球水冲突热点地区之一，其跨界水资源问题深刻影响着跨界水资源合作水平的提升、地区关系的稳定发展，以及中国西南周边的安全和中国与东南亚各国的区域合作布局。为妥善解决与湄公河沿岸国家的跨界水资源开发过程中产生的问题，降低开发成本，减少水资源投资风险，防止跨界水资源问题成为湄公河地区的战略风险点，中国需要展开高水平的水外交。

“水外交”一词初见于 1904 年，特指一切与水相关的外交活动，但学界至今尚未对此概念给出一个权威界定，每每论及跨界水资源管理和利用问题，仍沿用“跨界水资源合作”“外交中的国际河流事务”“水资源外交”等名词短语。

水外交实施主体多为本国政府及其有关部门，也包括企业和相关机构；

实施对象既包括域内外涉及跨界河流事务的国家，也包括与跨界水资源治理存在利益关联的相关行为体；实施途径基本涵盖所有外交途径，如外交谈判或交涉、外交访问、参加国际会议和国际组织、签署政府间协定或备忘录等。此外，作为国家总体外交不可或缺的组成部分，水外交还以水为核心在政治、经济、技术等诸多领域对外进行交流合作。凡此种种，均离不开能力建设。

能力建设有助于应对复杂的跨界水资源问题，是水外交活动的一项重要内容。作为中国周边外交的一部分，中国水外交侧重如何应对周边跨界水资源争端，在方式和手段上相较于欧美"水外交强国"显得单一，在水外交理论与实践研究及教育方面也有不足。因此，有必要发展社会、机构和个人的能力，如水资源管理能力、对国际水法和国际涉水机构的认知能力、争端解决能力、聚合各专业（如水文学、地质学、气候学、经济学、工程学、政治学、社会学、心理学、宗教学、文化学）人才的能力，特别是跨文化交际能力，从而充分发挥水外交作为中国与周边国家开展睦邻友好合作的重要平台作用。

本书语境中，跨文化交际能力既指本族语者与非本族语者之间的交际，亦指任何在语言、文化背景、专业背景方面有差异的人们之间的沟通交流。为加强水外交中跨文化交际能力的建设，本书尝试以水外交典型案例为切入点，将内容分为三大板块：概述部分（由南京师范大学区域国别研究院、南京师范大学中北学院谷俊仪撰写），上下游国家之间的水外交典型案例（即传统及常规水外交，由皖江工学院祝加贝撰写），非上下游国家之间的水外交典型案例（即广义水外交，由谷俊仪撰写）。框架由祝吉芳拟定。

写作此书，是希望水外交作为一种预防和化解跨界水资源争端、开展跨界水资源合作的新兴外交形式，能够吸引更多来自不同专业领域的年轻学子加入这项事业。

作者

2025 年 1 月

目录

Contents

Part Ⅰ

Introduction

Not speaking the same language is a universal complaint in intercultural and interdisciplinary collaboration. Indeed, whether in terms of a broader culture or ways of thinking, or the down-to-earth use of words, languages are key. So adopting a shared language is not a matter of convenience but rather a precondition for success. It is believed by many water professionals that water diplomatic efforts can succeed mainly through finding a common language to build joint visions of the issue of concern and mutual learning.

To try speaking the "same language" with people from other cultures or domains, Part Ⅰ is divided into three units as follows:

In Unit 1, we are to understand that global water problems caused by various reasons have prompted fears that water issues contain the seeds of violent conflict, but we are sure that a secure and sustainable water future can be ours if all the world's peoples work together.

Unit 2 is about the legal framework on the transboundary water cooperation. Much of water transcends state borders via shared river or lake basins or groundwater aquifers. The resulting political, economic, social and environmental interdependencies give water resources the crucial potential either to foster cooperation or exacerbate conflict. Despite the complexity of the problems, the last 50 years have seen two water conventions, more than 3,600 legal agreements and treaties signed and they have made international relations over water more stable and predictable.

Unit 3 deals with water diplomacy. Transboundary waters constitute a promising entry point for diplomats aiming for high peace dividends, warranting a strong interest from foreign policymakers. Yet, in the end, enhancing the governance of transboundary waters hinges on strengthening and connecting the institutions, national, international, and individuals that can channel political will into water cooperation on water resources.

To sum up, an increased awareness of serious water problems, more collaboration on transboundary waters within the national and international legal framework, and the rise of water diplomacy can help water diplomacy actors adopt a shared language, ignoring their differences in culture, values, national interests, etc. , although the common catchphrase "the problems are complex" will still be used by both scholars and practitioners for many years to come.

Unit 1 Global Water Problems

Background Study for Successful Intercultural Communication

Read authentic Chinese and English topical texts in advance before making a glossary for their associated terminology and jargon as follows.

water-related disasters 水灾害

water-related diseases, including cholera and schistosomiasis 介水传染病,如霍乱、血吸虫病

water-stressed countries 缺水国家

areas of high or extremely high water vulnerability 水脆弱性强或极强的地区

severe water scarcity 严重缺水

severely water-constrained agricultural areas 严重缺水的农业区

freshwater withdrawals 淡水抽取

irrigated areas 灌区

global irrigation water use 全球灌溉用水

global water demand (in water withdrawals) 全球需水量(取水量)

salinization of groundwater 地下水盐碱化

water quality data 水质数据

World Water Day held on 22 March every year 每年3月22日的“世界水日”

a limited water supply 有限的水供应

natural ecological space 自然生态空间

water resource depletion 水资源枯竭

black malodorous water bodies 黑臭水体

water resource conservation and recycling 水资源节约与循环利用

sewer networks in urban areas 城市污水管网

heavy-polluting sectors 重污染行业

solid wastes 固体废弃物

urban garbage disposal 城市垃圾处理
prioritizing water conservation 节水优先
major water conservation and water supply projects 节水供水重大水利工程
governance of major rivers 大河治理
reinforcement of diseased reservoirs 病险水库除险加固
prevention and control of mountain torrents 山洪灾害防治
flood control and disaster reduction 防洪减灾
disaster prevention and mitigation 防灾减灾
key river sections 重点河段
scarcity measurement 稀缺性评估

Text A
Water Scarcity

Pre-reading Questions

1. How would you view a friend of yours who is not mindful of his or her water usage?

2. What problems have you ever had in communicating with people who lack awareness of water conservation?

3. What do you think is responsible for water scarcity?

4. What kind of dangers would water scarcity bring us? And what solutions can be provided by individuals and governments?

5. How will you view the fact that economic water scarcity occurs when water is adequate, but unavailable due to lack of significant investment in water infrastructure?

Water scarcity is the state of shortage of water. It is a situation where sufficient water is not available to meet the demands of a specific place or region. According to the United Nations, a person needs a minimum of around 50 liters of water each day to meet the basic needs of drinking, cooking and hygiene. But the increasing population is failing this quantity of water and hence most of the population around the planet is affected by water scarcity, and many countries are mired in water shortage.

There are two key types of water scarcity: physical and economic. Physical scarcity occurs when the demand of the population exceeds the available water resources of a region. Symptoms of physical water scarcity are severe environmental degradation, declining groundwater, and water allocations that favor some groups over others. Arid regions are most often associated with physical water scarcity, but water scarcity also occurs in areas where water is apparently abundant, when water resources are overcommitted to various users owing to the overdevelopment of hydraulic infrastructure, most commonly for irrigation purposes. In the major rice-growing South Asian countries, per capita water availability decreased by 34% ~ 76% between 1950 and 2005 and is likely to decline by 18% ~ 88% by 2050. The water scarcity faced by this region is mainly economic, not physical.

Now water scarcity has become a serious global problem. But what caused this situation, then?

Globally, water can be scarce for many reasons: demand for water may be exceeding supply, climate change may be making water scarcity worse, water infrastructure may be inadequate, or institutions may be failing to balance everyone's needs.

The amount of water that can be physically accessed varies as supply and demand changes. As a relative concept, water scarcity intensifies as demand increases and/or as water supply is affected by decreasing quantity or quality. Water is a finite resource in growing demand. As the global population increases, resource-intensive economic development continues, rapid urbanization, changing lifestyles and consumption patterns significantly impact water requirements. Many countries' water resources and infrastructure are failing to meet accelerating demand. Growing demand for water increases the need for energy-intensive water pumping, transportation, and treatment, and has contributed to the degradation of critical water-dependent carbon sinks such as peatlands. Water-intensive agriculture for food production, particularly meat, and for growing crops used as biofuels, can further exacerbate water scarcity.

Global warming is coming to be one of the major causes of water scarcity. It is a condition where the average air temperature is becoming warmer and the water of rivers and lakes has been evaporating at a faster rate, which is leading to the drying up of many water bodies. Mainly due to the climate change, terrestrial water storage—the water held in soil, snow and ice—is diminishing. Meanwhile, glaciers, ice caps and snow fields are rapidly disappearing. Volatility in the cryosphere can affect the regulation of freshwater resources for vast numbers of people in lowland areas, for meltwater feeds many of the great river systems. Besides, flooding is becoming more frequent and more intense. Flooding can contaminate water resources with faecal matter, and cause damage to water and sanitation infrastructure such as water points, wells, toilets and wastewater treatment facilities. As sea-levels rise, toxins could concentrate in drying rivers, or saltwater could intrude into coastal groundwater aquifers.

U.S. space agency NASA's new analysis of 14 years of satellite data shows rapid change in the world's freshwater supply in startling detail never captured before. The 34 "trends" in the data picture not only the effects of climate change like worsening droughts, but also those of human over-use, such as pumping out underground aquifers to irrigate crops. Some 20 per cent of the world's aquifers are being overexploited. Stored in rocks and soil, groundwater is our biggest source of liquid freshwater. It sustains drinking water supplies, sanitation systems, farming, industry and ecosystems. But in many places, we simply do not know how much of this precious resource might exist.

Extreme weather events are making water even more scarce, more unpredictable, more polluted or all three. Some of the water reservoirs and hydropower stations should not be blamed for the events. In fact, the severe drought or flood has revealed that the water infrastructure along a river is "inadequate" to effectively fight droughts and floods. In the aftermath of disasters, infectious disease outbreaks are common due to the inadequate water infrastructure, resulting in the spread of sewage, the breakdown in water and sanitation services, and diminished ability to practise good hygiene.

It has been discovered that human destruction of vegetation and tree cover exacerbates soil erosion and reduces groundwater recharge, further increasing water scarcity.

Researchers have indicated where water scarcity is most likely to reach crisis mode and lead to armed conflict over resources and/or forced human migration. So it is essential to introduce some solutions to mitigate the crisis. Raising public awareness is crucial, and many social groups can be the facilitators to promote the importance of water conservation and impel the public to realize that this is not a national responsibility, but a civic duty. Besides, governments at all levels should help relevant regions to introduce advanced water-saving technologies and experience, such as upgrading the obsolete watering method into drip irrigation. What is more, the government can enact corresponding laws and regulations to stipulate and deter some water wasting behaviors. Overall, water scarcity is a global problem that requires the combined efforts of individuals and governments to solve.

Access to More Working Knowledge for Better Communication

1. Read the facts listed below and discuss in pairs or at a mock conference about how we can make a difference by trying to use a shared language and keep our intercultural communication focused around win-win.

—Only 0.5% of water on Earth is usable and available freshwater, and climate change is dangerously affecting that supply. Over the past 20 years, terrestrial water storage—including soil moisture, snow, and ice—has dropped at a rate of 1 cm per year, with major ramifications for water security.

—By 2050, the number of people at risk of floods will increase from its current level of 1.2 billion to 1.6 billion. In the early to mid-2010s, 1.9 billion people, or 27% of the global population, lived in potentially severely water-scarce areas. In 2050, this number will increase to 2.7 to 3.2 billion people.

—Over a fifth of the world's basins have recently experienced either rapid increase in their surface water area indicative of flooding, a growth in reservoirs and newly inundated land; or rapid declines in surface water area indicating drying up of lakes, reservoirs, wetlands, floodplains and seasonal water bodies.

—Water-related hazards have increased in frequency over the past 20 years. Since 2000, flood-related disasters have increased by 134%, and the number and duration of droughts also increased by 29%.

—21 million people, including 5 million children, live within 5 km of lakes with high turbidity (water cloudiness), which can indicate water pollution.

—Soil erosion from croplands carries away 25 ~ 40 billion tonnes of topsoil every year, significantly reducing crop yields and the soil's ability to regulate water, carbon and nutrients, and transporting 23 ~ 42 million tonnes of nitrogen and 15 ~ 26 million tonnes of phosphorus off land, with major negative effects on water quality.

—Naturally occurring arsenic pollution in groundwater now affects nearly 140 million people in 70 countries on all continents.

—While 45% of countries are on track to achieve their drinking-water coverage targets, only 25% of countries are on track to achieve their sanitation targets.

—42% of household wastewater is not treated properly, damaging ecosystems and human health.

—Between 2002 and 2021, droughts affected more than 1.4 billion people, causing the deaths of nearly 21,000 individuals.

2. Learn more about water scarcity by entering the following keywords before communicating with a friend who plays the role of a big polluter of our waterways, briefing him or her the water crisis and human need for access to clean water in a persuasive way.

UN World Water Development Report 2020: Water and Climate Change

United Nations Framework Convention on Climate Change (UNFCCC)/UN Climate Change: The Paris Agreement

UN-Water: UN-Water Policy Brief on Climate Change and Water 2019

UN-Water: World Water Day 2020 Campaign: Water and Climate Change

UN-Water: Summary Progress Update 2021: SDG 6—water and sanitation for all

Water and Climate Coalition: Water and Climate Leaders

The State of Climate Services 2021: Water Links

UN World Water Development Report 2020

UN International Strategy for Disaster Reduction (UNISDR) 2015

World Bank: Global Facility for Disaster Reduction and Recovery

World Bank: Water Resources Management

World Conference on Disaster Risk Reduction (WCDRR) 2015

The world's ecosystem services

Intergovernmental Technical Panel on Soils (ITPS) 2015

Intergovernmental Panel on Climate Change (IPCC): Climate Change and Water, 2008

Water balance of global aquifers revealed by groundwater footprint, 2012

Water Without Borders. Backgrounder. United Nations Department of Public Information, 2004

3. Search for more information on the following global water-related institutions from a variety of sources, verifying their achievements and calling for more individuals to join their efforts in conserving water and fighting against water pollution.

UN-Water 联合国水机制

World Bank 世界银行

World Health Organization(WHO) 世界卫生组织

Food and Agriculture Organization of the United Nations(FAO) 联合国粮食及农业组织

World Meteorological Organization(WMO) 世界气象组织

United Nations Office for Disaster Risk Reduction(UNDRR) 联合国减少灾害风险办公室

Development of Intercultural Communication Skills

1. Abundant metaphors about water have been created in English and Chinese. Summarize their major differences before manifesting what cultural factors can help lead to those differences.

2. Water is a culturally-loaded expression in almost all countries, which betrays the important role water plays in our daily life and communication. Do more selected readings and viewings (movies, TV documentary, historical books) to build a solid knowledge of the common cultural store of water in English and Chinese: cultural references, allusions, traditions, beliefs, history, folklore, children's stories and games and interpret more successfully between different and distant cultures.

3. Famous philosopher Laozi famously concludes that "the best goodness is like water". What does that suggest? How important is it for one in the Chinese

society to learn to behave appropriately? And how important is it for us when disputes over transboundary waters with downstream countries arise?

4. Think about these questions: Is water flowing downwards just a natural phenomenon in Chinese culture? Does it function the same as parental love, leniency, or unconditional spoiling in other cultures? If not, what are the differences?

5. Perhaps you have read about or witnessed behaviors (or non-verbal language) that separate us from our communicators in intercultural communication. Make a list and talk about them one by one.

Text B
Deteriorating Water Ecosystems

Pre-reading Questions

1. Empathy is one of the essential human traits, one that builds bonds between non-related individuals, encouraging sympathy between people who come from different cultures or different interest groups. Can we learn empathy? Do we communicate with someone who is different from us, like someone who is from an area of extremely high water stress?

2. Will you believe what someone is telling you about serious water problems in her hometown is true and then listen some more? Can you see what it's like for you to live in her shoes?

3. Could you please translate the sentence "He was confident that the Creator intended all men to be created equal"? Is it relevant to the so-called water rights and the demand for equal access to safe and adequate drinking water?

4. What do you think we can do to address the deteriorating global water problems?

An ecosystem is normally defined as a complex of all living (plants, animals, microorganisms) and non-living (soil, climate) components interacting as a functional unit in a certain area. Each contributes to maintaining the overall ecosystem's health and productivity. Life on earth depends on healthy ecosystems that can mitigate the effects of floods and droughts. Freshwater ecosystems, such as wetlands, rivers, mangroves, and aquifers, are a critical part of the global water cycle—supplying, purifying, and protecting freshwater resources. Coastal mangroves and river bank vegetation protect the land from erosion and inundation; wetlands and glaciers regulate water flows; and aquifers can act as a storage buffer during dry spells.

It is often conceptualized that ecosystems provide a range of "services" that can be categorized as: **provisioning** that refers to consumer goods, such as food and water; **regulating** that includes, among others, water purification and preventing erosion; **habitat** that provides the environment for life cycles of species or maintains genetic diversity, through quality and quantity of natural vegetation or substrate for fish and **cultural** that refers, for example, to the aesthetic, tourism and spiritual services.

It is widely acknowledged now that various ecosystems, both aquatic and terrestrial, are in decline, primarily due to economic development. There is no shortage of statistics. Since 1900, the world has lost around 50 per cent of its wetlands. Between $4. 3 and $20. 2 trillion per year worth of ecosystem services were lost between 1997 and 2011 due to land use change. An estimated 20 per cent of the world's aquifers are being over-exploited leading, among others, to land subsidence and saltwater intrusion. Over half of the world's large river systems are adversely affected by dams. Inefficient use of water for crop production has caused salinization of 20 per cent of the global irrigated land area. The decline of ecosystems results in a range of adverse impacts on humans, since billions of people live in water scarce regions, and/or areas with high water quality risks.

Mismanagement combined with climate change is devastating many ecosystems, undermining their ability to provide freshwater "services", which threatens the health of human societies and natural environments. Biodiversity in freshwater ecosystems is in danger. Freshwater species are going extinct more rapidly than terrestrial or marine species. Habitat loss and pollution are a growing threat.

Human impact on ecosystems is affecting water quantity and quality. Natural landscapes are being lost to urbanization, industry and agriculture. Climate change exacerbates the effect on water. Surface water area is changing fast. In some areas, flooding and increased rainfall are increasing surface water area. In other areas, lakes, wetlands and floodplains are drying up due to reduced precipitation and/or unsustainable management. Lake water high turbidity is a growing issue. High turbidity is where water appears "cloudy". The suspended particles act as magnets for pollutants and bacteria, affecting human and ecosystem health, and damaging tourism.

Industry and agriculture are often big water polluters. Increased usage of chemical fertilizers and pesticides, and untreated wastewater in irrigation, pollute groundwater and surface water. Industry in many areas still discharges waste directly into water courses.

Wastewater management is being seriously neglected. Across the world, safely reused wastewater is grossly undervalued as a potentially affordable and sustainable source of water, energy, nutrients and other recoverable materials. Untreated wastewater in cities is a major issue. In poorer urban areas, a large proportion of wastewater is discharged untreated directly into the closest drainage channel or water body. Household effluent, human waste, toxic chemicals and medical waste are exposed to the air, contaminating the environment in often densely-populated residential zones.

Ecosystem degradation is also a significant source of increasing water-related risks and extremes, such as floods and droughts. Ecosystems can provide natural ("green") infrastructure that can perform some disaster reduction functions, and therefore partially substitute or augment "grey" (built) infrastructure that targets the same purposes. Combining "green" and "grey" infrastructure, for example in the context of integrated flood and drought risk management in the same river basin, can lead to cost savings compared to grey infrastructure solutions alone. Also, green infrastructure provides functions and benefits that can directly improve the performance of grey infrastructure and help reduce risks to the latter. It is, however, unlikely that ecosystems alone can achieve the same risk reduction effect as grey infrastructure or totally replace it in the future. Therefore, advocating for ecosystems alone may be too simplistic for water-related disaster mitigation, and can potentially lead to ineffective policies.

Access to More Working Knowledge for Better Communication

1. Read the facts listed below and discuss in pairs or at a mock press conference about how to use our empathy to drive us to action.

—Only 11% of the estimated total of domestic and industrial wastewater produced is currently being reused.

—The untapped potential for wastewater reuse is around 320 billion cubic meters per year, with the potential to supply more than 10 times the current global desalination capacity.

—60% of water bodies assessed in 89 countries have good ambient water quality.

—Water quality data are not collected routinely in a majority of countries. This means that over 3 billion people are at risk because the health of their freshwater ecosystems is unknown.

—When a territory withdraws 25% or more of its renewable freshwater resources, it is said to be "water-stressed". Five out of eleven regions have water stress values above 25%, including two regions with high water stress and one with extreme water stress.

—Rising water stress is affecting food security and biodiversity. There are rapid changes in surface water in one fifth of river basins.

—Water-harvesting and water conservation techniques could boost rainfed kilocalorie production by up to 24%, and if combined with irrigation expansion, by more than 40%.

—The estimated economic value of ecosystem services globally, made in 2011, was $124.8 trillion, which was nearly twice the global gross domestic product in the same year.

—Increasing demands for water will substantially decrease the amount of water available for diluting wastes. Rapidly expanding industries which involve more and more complex chemical processes will produce large volumes of liquid wastes, and many of these will contain chemicals which are noxious.

—To feed our rapidly expanding population, agriculture will have to be intensified. This will involve ever-increasing quantities of agricultural chemicals.

—Power plant cooling is responsible for 43% of total freshwater withdrawals in Europe (more than 50% in several countries), nearly 50% in the USA, and more than 10% of the national water cap in China.

2. Learn more about global water problems by entering the following keywords before making an oral report on the deteriorating water ecosystems in your own words.

UN World Water Development Report 2024

UN Decade on Ecosystem Restoration

UN Environment Programme: UNEP Annual Report 2016

UN-Water: Progress on Water-related Ecosystems, 2021

International Water and Sanitation Centre (IRC): Financing universal access: the role of Water Financing Facilities, 2015

UN-Water: The Sustainable Development Goal 6 Global Acceleration Framework, 2020

UN-Water: Blueprint for Acceleration: Sustainable Development Goal 6 Synthesis Report on Water and Sanitation 2023

World Bank Group Water and Sanitation Program: The Costs of Meeting the 2030 Sustainable Development Goal Targets on Drinking Water, Sanitation, and Hygiene, 2016

World Health Organization (WHO): UN-Water Global Analysis and Assessment of Sanitation and Drinking-Water (GLAAS) 2022

World Bank Group: Overview of water

UN-Water: Progress on Ambient Water Quality—2021 Update

UN-Water: Summary Progress Update 2021: SDG 6—water and sanitation for all

UN World Water Development Report 2017: Wastewater: the untapped resource

United Nations Environment Programme (UNEP): Wastewater—Turning Problem to Solution

United Nations Environment Programme (UNEP): A Snapshot of the World's Water Quality: Towards a global assessment, 2016

Further Recommended Reading

[1] Nancy L. Peluso, and Michael Watts. *Violent environments*. New York: Cornell University Press, 2001.

[2] Peter H. Gleick. *The world's water 2008—2009*. London: Island Press, 2009.

[3] Michael R. Redclift, and Marco, Grasso. *Handbook on climate change and human security*. Cheltenham: Edward Elgar Publishing, 2013.

[4] Enamul Choudhury, Shafiqul Islam, Lawrence Susskind. *Complexity of transboundary water conflicts: enabling conditions for negotiating contingent resolutions*. London: Anthem Press, 2018.

Unit 2 Legal Framework on Water

Background Study for Successful Intercultural Communication

Read authentic Chinese and English topical texts in advance before making a glossary for their associated terminology and jargon as follows.

international watercourses 国际河流(侧重于国际水道的功能描述)

international rivers 国际河流(指河域位于多国边界内的河流)

international (drainage) basins 国际流域

shared river basins 跨境流域;国际流域(强调共享)

transboundary rivers 跨界河流(但不涵盖界河)

transboundary aquifers 跨界含水层

the upper, middle and lower reaches of the river 上、中、下游

the riparians 河岸国;流域国家或居民

co-riparians 跨界河流共有国;河流共有国

riparian nations 流域各国

the upper (or upstream) riparians 上游居民或国家

downstream riparians 下游居民或国家

upper basin riparians 流域内上游居民或毗邻国家

water events 水事件

water rights 水权(可以决定一国能否以其期望的方式使用水资源)

Mekong River Commission(MRC) 湄公河委员会

Sustainable Development Goals(SDGs) regions 可持续发展目标地区

boundary river; frontier river; international rivers as national borders 国际界河

the countries of the Mekong River Basin; the Mekong countries 湄公河流域国家

Beilun River, the boundary river between China and Vietnam 中越界河北仑河

the world's transboundary river basins 全球跨界河流流域

the global river discharge 全球河流流量

China's southbound transboundary river basins 中国南部跨界河流流域

downstream countries of the dam 大坝下游国家

lack of enforcement powers 没有执法权

reluctance to join as a full member 不愿作为正式成员加入

alternative institutional mechanisms 替代性体制机制

offering assistance for the construction of downstream dams 为下游大坝建设提供援助

absolute territorial sovereignty 绝对领土主权论

absolute territorial integrity 绝对领土完整论

limited territorial sovereignty 有限主权论

Text A
Conflict and Cooperation on Transboundary Water Resources

Pre-reading Questions

1. What makes China the main player in the regional water games?

2. Can you summarize the major differences between China and its neighboring states in transboundary water governance?

3. How should we look at water relations with our neighbors in the southeast? And what should we do to coexist harmoniously with them?

4. What do you think are the most effective ways to argue (or communicate) without rancor or faulty reasoning with co-riparians who have opposing viewpoints, with the hope of broadening our understanding of people and ideas?

Transboundary waters account for 60% of the world's freshwater flows and 153 countries have territory within at least one of the 286 transboundary river and lake basins and 592 transboundary aquifer systems, which often leads to disputes between upstream and downstream countries. Recently, Indian media reported that

Bhutan not only closed its border to prevent Indian border people from entering the country to take water, but also cut off the water flowing into India, leaving nearly 6,000 Indian residents inaccessible to water. The Indian media and people angrily questioned their foreign policy, and strongly asked their government to take a stronger attitude towards Bhutan without knowing the fact that Bhutan did not cut off the water supply intentionally. Due to the continuous heavy rain, its impoundment channel has been blocked, and the Bhutan government was stepping up sediment dredging.

For countries that share great rivers, allocation of water rights is a big problem. China and Kazakhstan had disputes over the development and utilization of the Ili River, but China has a larger share of the river, and around 12 to 15 billion cubic meters flow within its borders per year, almost twice that of Kazakhstan. Therefore, how to fairly distribute runoff is a problem for both the states, and reaching a satisfactory agreement on water allocation and quality control has posed a great challenge to the two states.

In modern history, water disputes have led to violence or even military actions. Yet these disputes are not likely to lead to "water wars". Of the almost 2,000 incidents that took place in transboundary basins between 1990 and 2008, approximately twice as many events were cooperative rather than conflictual. The Nile basin features significant conflict over access to and rights over the Nile water resources among its eleven riparian countries. The Nile Basin Initiative, founded by 10 out of 11 riparian countries in 1999 with backing from major donor institutions, has achieved some successes in its attempts to strengthen cooperation. Yet, since 2007, diverging interests between co-riparians have brought negotiations to a standstill, pitting Egypt against upstream riparians, especially Ethiopia. In 2015, trilateral negotiations between these countries over a major dam under construction in Ethiopia led to a framework agreement that may, in time, prepare the ground for a broader agreement. The Turkish-Armenian case is another prominent example of how two co-riparians can put their tensions aside, work together in their mutual interest, and share transboundary waters equitably.

While boundary-crossing water negotiations will always be difficult because competing interests are involved, it is imperative for states to cooperate over their transboundary waters, as reflected in the Sustainable Development Goals (SDGs):

"By 2030, implement integrated water resources management at all levels, including through transboundary cooperation as appropriate." The UN 2023 Water Conference called for the strengthening of transboundary water cooperation as a means by which to accelerate progress on sustainable development and regional integration, and to build sustainable peace.

Water can bring countries together and promote prosperity by providing common livelihood, development and cost-sharing opportunities that may exceed those generated by unilateral action. International water law has developed principles and norms that provide the basis for transboundary water cooperation, which can help solve disputes and contribute to stability and peace. The Convention on the Law of the Non-navigational Uses of International Watercourses (Watercourses Convention; United Nations, 1997) and the Convention on the Protection and Use of Transboundary Watercourses and International Lakes (Water Convention; UNECE, 2013), as well as the Draft Articles on the Law of Transboundary Aquifers (ILC, 2008), provide riparian states with basic norms for: equitable and reasonable utilization of the joint water body; the duty to do no significant harm; and the duty to cooperate. The UN water conventions include provisions for the peaceful settlement of disputes. To specifically prevent and resolve disputes, the Water Convention also has an implementation committee. In addition, regional integration bodies, such as the European Union and several regional economic communities across Africa, promote cooperation and peace through their coordination actions across sectors and river, lake, and aquifer systems.

Throughout history, countries have developed context-specific governance arrangements that foster peace and prosperity over transboundary waters. Research suggests that "coordination between stakeholders, through the establishment of institutional capacity in the form of agreements, treaties, or informal working relationships, can help reduce the likelihood of conflict. Once institutional capacity is established between parties, it has been proven to be resilient over time, even as conflict was being waged over other issues".

Currently transboundary waters globally face significant and increasing pressures due to population increase, growing water demands, ecosystem degradation, and climate change expected to increase the frequency and severity of extreme events like droughts and floods and degrade water quality. Cooperation over transboundary rivers, lakes and aquifers can decrease pressures by generating multiple economic, social, environmental and political benefits that in turn deliver prosperity and peace at local, national, regional and global levels. The topic has drawn attention at high levels at the United Nations, including through the Global High-Level Panel on Water and Peace which launched its first report in 2017. The UN Security Council has also recognized that joint water management can foster trust, stability and peace.

Access to More Working Knowledge for Better Communication

1. Read the facts listed below and discuss in pairs about how we can bridge the gap between the upstream and downstream riparians in intercultural encounters.

—In an era of increasing water stress, mismanaged transboundary water supplies have the potential to cause social unrest and spark conflict.

—The world's 286 transboundary lake and river basins cover nearly half of the Earth's land surface.

—A total of 145 nations include territory within transboundary lakes and river basins, and 30 countries lie entirely within such basins.

—While most basins are shared between just two countries, there are many basins where this number is much higher. There are 13 basins worldwide that are shared between 5 and 8 riparian nations. Five basins, the Congo, Niger, Nile, Rhine and Zambezi, are shared between 9 and 11 countries. The river that flows through the most nations is the Danube whose catchment area is shared by 19 countries.

—Worldwide there are approximately 300 transboundary aquifers and 153 countries with transboundary water bodies. These figures mean that over 90% of the world's population lives in countries that share basins, and about 40% of the world's population is found in transboundary basins.

—Since 1948, 37 incidents of acute conflict over water have happened, 295 international water agreements have been signed, including the UNECE Water

Convention, and 116 river basin organisations have been established reflecting differing levels of transboundary cooperation.

—Although the number of transboundary water disputes in the last 70 years can be seen as low, much work is needed in order to reach peaceful and operational agreements.

—Only 24 countries report that all their transboundary basins are covered by cooperation arrangements.

—Only 32 countries have 90% or more of their transboundary basin area covered by operational arrangements.

—Only one SDG region is on track to have all its transboundary rivers, lakes and aquifers covered by cooperation arrangements by 2030.

—Transboundary water cooperation is not attracting enough investment. Financial constraints and limited understanding among governments of the benefits of cooperation on water resources often hinder the mobilization of financial resources for transboundary water cooperation. The perceived high risk due to lack of formal arrangements is off-putting to investors in many cases.

—Ineffective transboundary water management will worsen the increasingly serious transboundary water problems such as unfair division, over-use, pollution, failure to recycle, failure to conserve, floods and droughts which have been leading to water conflicts and even wars. As water quality degrades or the quantity available has to meet rising demands over time, competition among water users intensifies. This is nowhere more destabilizing than in river basins that cross political boundaries.

—The shared waters of transboundary basins are a source of both conflict and opportunity for cooperative behavior. Increased variability and uncertainty related to water quantity and quality may multiply both water-related risk and opportunity for riparian states sharing transboundary waters.

—To harvest the benefits of water cooperation, we have to go beyond the absence of conflict and do things jointly.

2. Find more information about the following famous cases of water conflicts, from which we can look into how water and conflict overlap.

droughts, livestock prices and armed conflict in Somalia 索马里的干旱、牲畜价格和武装冲突

security implications of growing water scarcity in Egypt 埃及日益缺水对安全的影响

water privatisation in Cochabamba, Bolivia 玻利维亚科恰班巴的水私有化

dispute over water in the Cauvery Basin in India 印度高韦里河流域水资源争端

conflict over the Euphrates-Tigris of Turkey, Syria and Iraq 土耳其、叙利亚和伊拉克之间的幼发拉底河-底格里斯河冲突

disputes over water in the Nile Basin 尼罗河流域水资源争端

water shortages and public discontent in Yemen 也门的水资源短缺与公众不满

transboundary water disputes between Afghanistan and Iran 阿富汗和伊朗之间的跨界水争端

disputes over water of India with other countries 印度与多国水资源之争

3. Search for more information about the cases of successful water cooperation listed below and elaborate on how important identifying a common concern is in negotiating a water agreement against all odds to bridge conflicting water interests.

water cooperation despite tensions in Turkey and Armenia 土耳其、亚美尼亚搁置仇隙展开水资源合作

Indus Waters Treaty between India and Pakistan (IWT) 印度、巴基斯坦签署的《印度河水域条约》

Nile Basin Initiative(NBI) 尼罗河流域倡议

Nile Basin Cooperative Framework Agreement《尼罗河流域合作框架协议》

Ganges Water Treaty between Bangladesh and India(GWT) 孟加拉国、印度签署的《恒河流域条约》

Israel-Jordan Peace Treaty《以色列-约旦和平条约》

Ayeyawady-Chao Phraya-Mekong Economic Cooperation Strategy (ACMECS) 伊洛瓦底江-湄南河-湄公河三河流域经济合作战略机制(三河流域机制)

Greater Mekong Subregion Economic Cooperation Program(GMS) 大湄公河次区域经济合作机制

LMC Special Fund 澜湄合作专项基金

Water Resources Cooperation Center, the Environmental Cooperation Center and Global Center for Mekong River Studies 水资源、环境合作中心和全球湄公

河研究中心

Lancang-Mekong Cooperation Leaders' Meeting 澜湄合作领导人会议

Transboundary Freshwater Dispute Database (TFDD) 跨界淡水争端数据库

transboundary institutions for specific water systems 特定水系的跨界机构

Global Water Partnership 全球水伙伴关系

Organization of the Senegal Riparian States 塞内加尔河流域国家组织

Intergovernmental Coordinating Committee of the River Plate Basin 拉普拉特河流域国家政府间协调委员会

Transboundary Waters Assessment Programme (TWAP) 跨界水域评估计划

Text B
Transboundary Water Conventions and Agreements

Pre-reading Questions

1. Defined as "the ability to understand and share the feelings of another", empathy is at least part of the reason human beings have been so successful as a species. Can it work across contending co-riparians, encouraging sympathy between different riparian states with competing interests?

2. Water rights are based on the idea that people have access to basic sanitation and that everybody should be provided access to safe drinking water. What do you think may interfere with our international efforts to reach consensus on water rights and nations' mitigation obligations?

3. Should the riparians act in accordance with their interests or the law? Is there any legally binding international law for riparian nations to obey when disputes over transboundary waters arise?

4. There are two legal framework instruments covering transboundary water issues in force at the global level: the UN Watercourses Convention and the Water Convention. Can the two global UN water conventions effectively coexist? And how could they be implemented in a coordinated manner?

As water quality degrades or the quantity available has to meet rising demands over time, competition among water users intensifies. This is nowhere more destabilizing than in river basins that cross political boundaries. Despite the complexity of the problems, records show that water disputes can be handled diplomatically. The last 50 years have seen only 37 acute disputes involving violence, compared to 150 treaties that have been signed. Nations value these agreements because they make international relations over water more stable and predictable. In fact, the history of international water treaties dates as far back as 2500 BC, when the two Sumerian city-states of Lagash and Umma crafted an agreement ending a water dispute along the Tigris River—often said to be the first treaty of any kind. Since then, a large body of water treaties has emerged. According to the Food and Agriculture Organization, more than 3,600 treaties related to international water resources have been drawn up since 805 AD. The majority of these deal with navigation and boundary demarcation. The focus of negotiation and treaty-making in the last century has shifted away from navigation towards the use, development, protection, conservation, and sharing of water resources.

Legal agreements on water sharing have been negotiated and maintained even as conflicts have persisted over other issues. Cambodia, Laos, Thailand and Vietnam have been able to cooperate since 1957 within the framework of the Mekong River Commission, and they had technical exchanges throughout the Vietnam War. Since 1955 Israel and Jordan have held regular talks on the sharing of the Jordan River, even as they were until recently in a legal state of war. The Indus River Commission survived two wars between India and Pakistan. A framework for the Nile River Basin, home to 160 million people and shared among 10 countries, was agreed in February 1999 in order to fight poverty and spur economic development in the region by promoting equitable use of, and benefits from, common water resources. The nine Niger River Basin countries have agreed on a framework for a similar partnership. These cases reflect two important elements of international water resources cooperation: the need for an institution to effectively develop a process of engagement over time; and well-funded third-party support trusted by all factions.

The more than 3,600 agreements and treaties signed are an achievement in themselves, but a closer look at them still reveals significant weaknesses. What are needed are workable monitoring provisions, enforcement mechanisms, and specific water allocation provisions that address variations in water flow and changing needs. The Convention on the Protection and Use of Transboundary Watercourses and International Lakes(Water Convention) was adopted in 1992 and entered into force in 1996. It brings together almost all countries sharing transboundary waters in the pan-European region, and is expected to achieve broader participation with its global opening to all United Nations Member States. The 1997 United Nations Convention on Non-navigational Uses of International Watercourses, as one international instrument that specifically focuses on shared water resources, established two key principles to guide the conduct of nations regarding shared watercourses: "equitable and reasonable use" and "the obligation not to cause significant harm" to neighbours. However, it is up to countries themselves to spell out precisely what these terms mean in their watersheds.

The UN International Law Commission(UNILC) embarked on the codification of the law of transboundary aquifers in 2002 in order to provide a legal regime for the proper management of aquifers. The 63rd session of the UN General Assembly adopted Resolution A/RES/63/124 on the Law of Transboundary Aquifers by consensus on 11 December 2008. The resolution encourages states to make appropriate bilateral or regional arrangements for the proper management of their transboundary aquifers, taking into account the provisions of these draft articles, which are annexed to the resolution. These provisions include cooperation among states to prevent, reduce and control pollution of shared aquifers. In view of the importance of these "invisible resources", states are invited to consider these draft articles as a basis for the elaboration of a convention. The Law of Transboundary Aquifers is a concrete step forward towards the peaceful sharing of groundwater resources. Until then there was no instrument of international law that could provide a complete set of recommendations and guidelines for the sustainable and peaceful management of transboundary aquifers.

However, water professionals are still of the opinion that international water agreements or laws need to be more concrete, setting out measures to enforce treaties made and incorporating detailed conflict resolution mechanisms in case disputes erupt. Better cooperation also entails identifying clear yet flexible water allocations and water quality standards, taking into account hydrological events, changing basin dynamics, and societal values.

Access to More Working Knowledge for Better Communication

1. Read the facts listed below and discuss in pairs or at mock meetings about how an intercultural approach could be efficiently applied to make international relations over water more stable and predictable.

—Preventing conflicts from arising in the first place is far more efficient than interventions at a later stage.

—One of the ways that countries have managed water issues is through commissions and conventions.

—In March 1992, governments gathered in Helsinki, Finland, with a vision in mind: to manage shared waters in the pan-European region collaboratively and sustainably. By the end of the meeting, the Water Convention was born. Today, this Convention is an effective framework for advancing transboundary water cooperation at the global level.

—International Freshwater Resources Law includes instruments adopted at the universal level, such as the United Nations Convention on the Law of the Non-navigational Uses of International Watercourses and the International Law Commission's Draft Articles on the Law of Transboundary Aquifers and laws at the regional and basin levels.

—Freshwater and International Law focuses on the interplay between universal, regional and basin perspectives.

—There are multilateral institutions for the management of international river basins, lakes, and aquifers that define the scope of authority, internal structure, decision-making procedures, and dispute settlement mechanisms.

—International Water Law plays an important role in the promotion of cooperation over shared transboundary watercourses.

—There are more than 300 international freshwater agreements, and a collection of thematic maps related to the agreements, their content, and the river basins they represent.

—In February 2011, the UN General Assembly decided to proclaim 2013 International Year of Water Cooperation.

—The 1997 United Nations Convention on the Law of the Non-navigational Uses of International Watercourses is the only treaty governing shared freshwater resources that is of universal applicability. It is a framework convention, in the sense that it provides a framework of principles and rules that may be applied and adjusted to suit the characteristics of particular international watercourses. Some key guiding principles set out in the document include: the equitable and reasonable utilization of international watercourses; the application of appropriate measures to prevent harm to other states sharing an international watercourse; and the principle of prior notification of planned measures. For the Convention to become legally binding, at least 35 nations must ratify it.

—World Water Day 2009 was dedicated to the theme "Shared Waters, Shared Opportunities". The focus was placed on transboundary waters and on the fact that nurturing the opportunities for cooperation in transboundary water management can help build mutual respect, understanding and trust among countries and promote peace, security and sustainable economic growth.

—The UN-Water Thematic Priority Area (TPA) on Transboundary Waters is intended to provide a platform to promote coherence and coordination of activities by UN-Water Members and Partners in the area of transboundary waters. It does this by facilitating a steady exchange of information, experiences and lessons learned and by promoting joint efforts.

—The UNECE Convention on the Protection and Use of Transboundary Watercourses and International Lakes (the Water Convention), adopted in 1992 and entered into force in 1996, serves as a mechanism to strengthen international cooperation and national measures for the ecologically sound management and protection of transboundary surface waters and groundwaters. Furthermore, it provides an intergovernmental platform for the day-to-day development of transboundary cooperation. The Convention obliges riparian parties to prevent, control and reduce transboundary impact, use transboundary waters in a reasonable and equitable way and ensure their sustainable management. Parties bordering the same transboundary waters shall cooperate by entering into specific agreements and establishing joint bodies. The Convention includes provisions on monitoring, research and development, consultations, warning and alarm systems, mutual assistance, and exchange of information, as well as access to information by the public.

2. Read the paragraph below and elaborate on how language barriers can be overcome in solving transboundary water problems.

To be clear, language barriers may exist in theory-driven collaboration as they do in problem-driven work. However, in theory-driven work, collaborators can fall back on the newly evolving vocabulary that belongs to their unified theory. In lieu of this unified theory, we advocate that problem-driven collaborations explicitly dedicate time to create, adapt, or adopt a shared framework for understanding, describing, and explaining the problems they will be addressing. One such framework is the Water Diplomacy Framework (WDF). The WDF was originally introduced in 2013 and has co-evolved throughout the process of adopting and adapting it in a problem-driven interdisciplinary research setting.

The WDF is far from a general theory for solving complex water issues. Instead, it provides a shared vocabulary of common concepts encountered when confronting problems arising from the complexity of coupled natural and human systems.

We see this not just in our own experience but in frameworks such as the International Classification of Functioning, Disability and Health (ICF), which was established by the World Health Organization as an interprofessional language with which to communicate health information across disciplinary and geographic boundaries. Analogous to the WDF, the ICF provides a common language, terms and concepts, a suite of strategies to facilitate cooperation and integrate different perspectives, and a set of inviolable principles for professional practice. It can be seen as a meta-language to help clarify the relationship between data, information and knowledge, and to build a shared understanding and interpretation of concepts. Indeed, the ICF has received a warm welcome because the provision of a shared framework and language are fundamental steps in achieving the goal of interprofessional collaboration. While the WDF has not been formalized to the level of the ICF, we too hope that as the framework matures, it can meet the ICF's goal of being intelligible for all stakeholders.

3. Switzerland is considered to be the "chateau of water" in Europe, with six water sources and four lakes that the country shares with its neighbours. The country's first transboundary agreements are the result of complex negotiations and compromise, highlighting benefits of transboundary water management. Detail concrete experiences of successful transboundary cooperation based on the following list of international and national water-related laws and regulations.

the Transboundary Waters Assessment Programme (TWAP)《跨界水域评估计划》

EU Water Framework Directive《欧盟水框架指令》

Council Conclusions on EU Water Diplomacy《关于欧盟水外交的理事会决议》

regional conventions 区域性公约

Water Diplomacy Framework《水外交框架》

the United Nations European Commission of Economy(UNECE) Convention on the Protection and Use of Transboundary Watercourses and International Lakes (Water Convention) 联合国欧洲经济委员会《跨界水道和国际湖泊保护和利用公约》(《水公约》)

the 2003 Amendment to the Water Convention 2003 年《水公约》修正案

the United Nations Convention on the Law of the Non-navigational Uses of International Watercourses (Watercourses Convention)《联合国国际水道非航行使用法公约》(《水道公约》)

the Law of Transboundary Aquifers《跨界含水层法》

International Water Law《国际水法》

the Global Water Partnership(GWP) 全球水伙伴

The South Asia Water Initiative (SAWI)《南亚水资源倡议》

Lancang-Mekong Agreement《澜湄协定》

Water Law of the People's Republic of China《中华人民共和国水法》

Environmental Protection Law of the People's Republic of China《中华人民共和国环境保护法》

Water Pollution Prevention and Control Law of the People's Republic of China《中华人民共和国水污染防治法》

Soil and Water Conservation Law of the People's Republic of China《中华人

民共和国水土保持法》

4. Discover the latest news on what is happening globally, exploring newly released publications on transboundary water cooperation from sources or links such as:

- Intergovernmental Panel on Climate Change (IPCC) Sixth Assessment Report: Fact Sheet—Food and Water
- Intergovernmental Panel on Climate Change (IPCC) Climate Change 2014: Impacts, Adaptation, and Vulnerability
- United Nations Children's Fund (UNICEF) Water under fire: the role of water in conflicts around the world
- United Nations Children's Fund (UNICEF) Children living in protracted conflicts are three times more likely to die from water-related diseases than from violence
- United Nations Educational, Scientific and Cultural Organization (UNESCO) Transboundary aquifers: challenges and the way forward
- UN-Water Blueprint for Acceleration: Sustainable Development Goal 6 Synthesis Report on Water and Sanitation 2023
- UN-Water Summary Progress Update 2021: SDG 6—water and sanitation for all
- UN-Water GLAAS 2014: Investing in Water and Sanitation: Increasing Access, Reducing Inequalities
- World Meteorological Organization (WMO) Protect our people and future generations: Water and Climate Leaders call for urgent action
- World Meteorological Organization (WMO) 2021 State of Climate Services: Water
- World Bank Water resources management
- UN Documentation Centre on Water and Sanitation (UN DCWS)
- the 2030 Agenda for Sustainable Development
- http://www.uswaterpartnership.org
- https://www.unwater.org/news/rome-water-dialogue
- https://www.un.org/waterforlifedecade/water_cooperation_2013/water_convention.shtml
- https://www.unece.org/environmental-policy/environmental-monitoring-

and-assessment/areas-of-work/

- https://www.internationalwaterlaw.org
- https://waterdiplomacy.org

Further Recommended Reading

[1] Seth M. Siegel. *Let there be water: Israel's solution for a water-starved world.* New York: Thomas Dunne Books, 2015.

[2] Philip Ball. *The water kingdom: a secret history of China.* London: Bodley Head, 2016.

[3] Stephen McCaffrey. *The law of international watercourses: non-navigational uses Oxford monographs in international law.* Oxford: Oxford University Press, 2003.

[4] Susanne Schmeier. *Governing international watercourses: river basin organizations and the sustainable governance of internationally shared rivers and lakes.* New York: Routledge, 2013.

[5] United Nations Economic Commission for Europe. *Frequently asked questions on the 1992 water convention with the road map to facilitate accession processes.* New York: United Nations Publication, 2020. https://doi.org/10.18356/9789210051262.

Unit 3　Rise of Water Diplomacy

Background Study for Successful Intercultural Communication

Read authentic Chinese and English topical texts in advance before making a glossary for their associated terminology and jargon as follows.

unfair division of water resources 水资源划分不公

unfair allocation of water resources 水资源分配不公

over-use of groundwater 地下水过度使用

upstream pollution sources 上游污染源

downstream solid waste disposal 下游固废处理

illegal sewage discharge from mining enterprises 矿业违法排污

discharge sewage without treatment or disposal 排放未经处理的污水

failure to recycle 未能回收

failure to conserve water 未能节约用水

losing water fast 失水快

common water claims 常见的水索赔

treatment facilities 污水处理设施

periodic droughts and forest fires 周期性干旱和森林火灾

future water infrastructure 未来水利基础设施

hydrologic extremes 极端水文事件

sustainability of the water quality 水质的可持续性

framing mutual interests through joint fact finding 通过联合实况调查确定共同利益

interdependence among contending stakeholders 相互竞争的利益相关者之间的相互依存关系

the hydro-political relationships between riparian states 沿岸国之间的水政治关系

projects in transboundary watersheds or other multi-stakeholder water systems 跨界流域或其他多方利益攸关方供水系统项目

competing interests 相互冲突的利益

bilateral tensions over decreasing flow rates 水流量变小引发的双边紧张局势

quick fix to the water contention 快速解决争水问题的办法

public decisions in shared waters 共享水域的公共决策

the joint management of aquifers, lakes and river basins 含水层、湖泊和河流流域的联合管理

non-water-related benefits 非涉水利益

water quality monitoring 水质监测

reclaim wastewater for agriculture and desalination of seawater 回收废水用于农业和淡化海水

groundwater rights in exchange for the supply of desalinated water 用地下水权利换淡水供应

Text A

Bringing Diplomacy into Solution to Water Problems & Water into Diplomacy

Pre-reading Questions

1. What does diplomacy have to do with water? On the face of it, perhaps not much. But why water diplomacy? Is cooperation over transboundary surface and ground waters the exclusive domain of water diplomats and foreign policy experts? Why should non-governmental stakeholders be involved in transboundary water dialogues?

2. What lies at the heart of debates and dialogues around the theory and practice of water diplomacy?

3. Oil and water are naturally incompatible, but can people who have nothing in common get along well? How would a water professional communicate more efficiently with his or her counterparts who declare competing interests?

4. A Fudan University professor hailed water diplomacy in 2021 as a new

path for China to build soft power in the Mekong River. Are you for or against this view? And why?

Diplomacy, a multifaceted and dynamic discipline that requires a profound understanding of foreign cultures, traditions, and politics, generally refers to the profession, activity, or skill of managing international relations, typically by a country's representatives abroad. Its synonyms are statesmanship and statecraft. It also means the art of dealing with people in a sensitive and effective way. It is mainly about negotiation, persuasion, and the art of finding common ground in the midst of differing interests and beliefs. Simply put, diplomacy is to meet your interests very well while meeting the interests of others reasonably well.

As an ancient practice as old as civilization itself, diplomacy has always been a crucial tool in the quest for peace and understanding among nations. The art of negotiation and persuasion, exercised by diplomats throughout history, has often been the difference between war and peace, conflict and cooperation. But it is not only about the exchange of words but also about understanding the nuances of culture, traditions, and the political landscape. Diplomats are often required to possess a profound knowledge of foreign languages, customs, and politics. They are expected to be adept at reading between the lines, understanding the unspoken messages and subtle cues, which are particularly true in the interconnected world, where the principles of jurisdiction, rule of law, and the intrigue of party politics often play a pivotal role in shaping diplomatic outcomes and ensuring justice. In ancient China, for instance, diplomats were highly regarded for their wit, intelligence, and diplomatic skills. They were often sent abroad to negotiate treaties, seal alliances, and resolve conflicts. These diplomats, known as "shi" or "diplomatic envoys", were expected to exhibit the utmost courtesy and respect while engaging with foreign dignitaries. They were also required to be conversant with the intricacies of foreign policy and international law, ensuring that their actions and words were in line with the interests and dignity of their country. The ancient Greeks and Romans, on the other hand, emphasized the role of rhetoric and persuasion in diplomacy. Statesmen like Caesar and Cicero were renowned for their eloquence and ability to sway public opinion through powerful speeches and debates. These diplomats understood that diplomacy was not just about negotiation but also about shaping public perception and influencing decision-making processes within and beyond their own borders.

Like ancient diplomacy, today's diplomacy is not always smooth sailing. The complexities of international relations and the differing interests of competing states often lead to tense situations and outright conflicts. In such cases, diplomats are required to exercise extreme caution and skills, often finding themselves in the middle of intricate intrigues and power games. Despite these challenges, they persevere, striving to find common ground and establish lasting peace between the warring or conflicting states. Some of the strategies and techniques employed by these diplomats are also often used by a very special group of people popularly known as water professionals and officials in addressing transboundary water problems, serving as a valuable guide for them as they navigate the complex waters of global or regional disputes over boundary-crossing waters.

Bringing diplomacy into solution to water problems and water into diplomacy, water diplomacy(WD, also known as hydro-diplomacy) may be understood as the measures that may be taken to prevent or peacefully resolve conflicts over water resources concerning their availability, allocation or use between and within countries. The concept is preventive in nature as well as offering an approach for conflict resolution. Often viewed as a tool to be deployed when conflicts over water use or the development of a basin surface, potentially impacting the relations of the countries concerned beyond water management, water diplomacy could, perhaps, to a greater degree, take inspiration and seek solutions from established practices and frameworks where functional cooperation has been established and major differences have been avoided over the management of shared waters. This should be held with the understanding that cooperation practices have developed in particular local contexts. The key features of cooperation arrangements that have brought about sustained cooperation, dialogue and the resolution of various issues, and that have also been adapted to deal with emerging challenges, merit attention.

Water diplomacy usually deals with competing interests locally, nationally, regionally and internationally. The four levels of water diplomacy can and should learn from each other to apply successful techniques of clear communication, transparency, data sharing and alignment, productive debate, and how to make fair but tough choices; all critical for effective diplomacy. The obligation to establish agreements on shared waters and joint bodies to govern them has contributed to gathering important experience, and some of this has been synthesized into the Principles for Effective Joint Bodies which were adopted by the Meeting of the Parties to the UNECE Water Convention in 2015. The Principles are intended as a resource to increase the effectiveness of joint bodies in transboundary water cooperation.

Rivers, tributaries, lakes and aquifers know no borders. Over the years, water management has more often been a source of cooperation than one of confrontation. As we look towards a future where water cooperation and understanding are paramount, it is worth borrowing the wisdom and insights of those who came before us, who through their diplomatic efforts, have paved the way for a more peaceful and understanding world. Yet, in the end, governance of transboundary waters hinges on strengthening and connecting the institutions, national and international, that can channel political will into coherent action.

Access to More Working Knowledge for Better Communication

1. Read the facts listed below and discuss in pairs or at a mock negotiating session about how we might make a difference by trying to bridge the gap between contending riparian states.

—Even in the past 50 years, one-fourth of water-related interactions have been hostile, ranging from name calling to military action.

—Climate change, geopolitical unrest, pandemics, mass migration, hyperinflation and other crises exacerbate water access inequalities. In nearly all cases, the poorest and most vulnerable groups are those that suffer the greatest risks to their well-being.

—A dam could generate hydroelectricity, but failing to control the deforestation upstream could result in rampant soil erosion that gets deposited in the water.

—Water security leads to prosperity and peace, while the hardships of conflict are amplified through water.

—Water diplomacy is a theory and practice of implementing adaptive water management for complex water issues, developed at Tufts, MIT, and Harvard.

—The water diplomacy approach diagnoses water problems, identifies intervention points, and proposes sustainable solutions that are sensitive to diverse viewpoints and values, ambiguity and uncertainty, as well as changing and competing needs.

—Increasingly complex water problems require negotiated solutions.

—Water diplomacy teaches the skills to transform a fixed quantity of water into a flexible and sustainably shared resource.

—Every year, UN-Water—the UN's coordination mechanism on water and sanitation—sets the theme for World Water Day. In 2023, the focus was on "Accelerating Change". In 2024, the theme was "Leveraging Water for Peace".

—Water diplomacy is a novel framework within the developing field of interdisciplinary water management.

—50% of jobs in high-income countries depend on water, a figure that increases to 80% in the low-income countries.

—In an attempt to address these complex water issues, the Water Diplomacy Framework (WDF) diagnoses the sources of complexity for a given water problem, identifies intervention points, and proposes sustainable resolutions that are sensitive to diverse viewpoints and uncertainty as well as to changing and competing needs.

—The WDF actively seeks value creation opportunities by blending science, engineering, policy and politics through a contingent negotiated approach.

2. Read more books and papers on water diplomacy and talk about the following arguments, discussing whether you are for or against them and why.

—Water diplomacy, as a multi-disciplinary concept, is located at the intersection of science, policy, and practice, assuming participation from all these domains.

—Water diplomacy links foreign policy and diplomacy with shared waters.

—Water diplomacy is an approach for peace mediation, as well as conflict prevention and mitigation.

—Water diplomacy shows 5 aspects: political, preventive, integrative, cooperative, technical.

—Water diplomacy is an emerging concept to address the political nature of transboundary cooperation and to link water with broader regional cooperation, geopolitics and foreign policy.

—Transboundary water cooperation and water diplomacy are distinct concepts.

—Water diplomacy is gaining increasing attention among both researchers and policy-makers. The interest is understandable, given that the concept brings together themes such as shifting geopolitics, new types of diplomacy, and increasing water scarcity.

—Actual water diplomacy actions typically vary across multiple tracks and scales.

—Water diplomacy, as a concept and as a practical approach, provides an example of the future of foreign policy and diplomacy, where the use of shared waters is likely to be of increasing importance.

—The concepts of water diplomacy and transboundary water cooperation are at times even used interchangeably, although water diplomacy studies typically focus more on conflict prevention and diplomatic engagement.

—The close connection and overlaps that water diplomacy has with transboundary water cooperation may create additional confusion, particularly among actors who are external to the water field.

—Water cooperation can be seen as an operative end-product of the process of water diplomacy, with the very aim of the cooperation defined by the diplomatic process.

—Water diplomacy and transboundary water cooperation are closely connected but separate concepts.

—Water diplomacy goes beyond the conventional regional treaties and transboundary cooperation mechanisms between riparian country governments, involving also other actors and themes.

—Instruments of international water law, among them notably conventions and soft law developed on that basis—which promote equitable and reasonable use of transboundary waters and prevention of transboundary impact—can serve as tools for water diplomacy.

Development of Intercultural Communication Skills

1. We sometimes use the saying "oil and water do not mix" to express an opinion that two things do not go together. How should such a situation be handled by a would-be water diplomat in his or her training?

2. If you pour oil on troubled water, you can calm a tense situation. How would you be able to become the oil on our troubled waters if you were smeared and branded as a water manipulator?

3. Usually, it's ethnocentrism that leads to failure in water diplomacy. Do you know what the word "ethnocentrism" means? How should we confront ethnocentrism in future negotiations on transboundary water issues?

4. Some people believe that being against your counterparts or something is a prejudice. What do you say to it? What do you think are the main causes of prejudice? And what should we do to overcome the prejudice in water negotiations and collaboration?

5. The current world is a complete tapestry of diverse cultures, each with its own unique systems of water governance and diplomacy. How shall we overcome the cultural differences in transboundary water cooperation?

6. The Nile, Mekong, and Danube are notable examples, and most of these agreements or treaties seek to resolve or avoid conflict over shared water use. While these fall under the umbrella of "Water Diplomacy", we should be careful not to restrict water diplomacy to negotiations between foreign governments, since contentious water use agreements are debated intrastate, between communities, between industries, and even between individual water users. Identify and summarize the role of explicitness in communication like water negotiations and debates.

7. Summarize the risks of water disputes between China and the Mekong River Basin, and propose China's corresponding innovative water diplomacy strategies.

8. Try using public data to scientifically interpret the distribution of water resources in various countries in the Lancang-Mekong River Basin, effectively

fighting back the false statements of European and American countries about China's upstream water conservancy facilities causing water shortages in the lower Mekong River.

9. Do some research on the evolution of water diplomacy, the theoretical construction of water diplomacy, China's water diplomacy practice in the Mekong River since 1980, and the risks China faces in water diplomacy and innovative strategies to it.

Text B
Water Diplomacy Actors

Pre-reading Questions

1. What skills do water professionals need to be effective in situations that require working across boundaries? What self-understanding do they need in order to conceptualize their role as an expert engaged in inclusive fact-value conversations and the co-production of knowledge with stakeholders?

2. Being able to experience or understand someone else's viewpoint is an incredible way to see what it is like to live in someone else's shoes. Is there any better tool, do you think, for helping you appreciate your own situation or to help you change your mind about the unequal distribution of water resources?

3. Should water negotiators only act on an invitation from all basin states, from one or a group of riparians or on their own initiative?

4. Would you please take the practice of water diplomacy between China and the Lancang-Mekong River Basin countries to resolve water disputes as a case study before identifying actors in water diplomacy?

There has historically been more cooperation than conflict over transboundary waters. But given changes brought on by climate change and growing human demands for water, disputes will arise unless shared water resources are managed through international law, science, and mutual gains-based water diplomacy.

The term water diplomacy refers to processes led by state actors and their representatives to advance national interests related to the management of shared water resources. State actors have the formal obligation to respect, protect, and fulfill their citizens' right to water, while other official actors of international public law (i. e., intergovernmental organizations) can also be involved in providing third-party support such as good offices or mediation.

The term is also broadly understood as a process engaged in by non-state actors to address, resolve, or avoid conflicts over access to water and water-related issues. It can represent a skill set possessed by an individual or a group of individuals(e. g., mediators) exercised in dealing with conflicts over water on behalf of state or non-state actors.

As a matter of fact, most water issues have a historic and well-defined set of actors: civil engineers, hydrologists, regulators, academics, civil servants, policy makers, NGOs, and politicians, ranging from international, regional and national governmental and intergovernmental institutions to representatives of indigenous tribes living on opposite shores of a river, including non-official representatives of a given state, local and spiritual leaders, grassroots groups, opponents to a state's position, and any kind of non-governmental organization. In the Columbia Basin example, public participation successes have been a keystone of the diplomatic process. The societal capacities manifested in the public participation refer to the competencies of our communities to engage in decision-making processes that impact them, and also the competencies of political leadership to seek, listen to, and incorporate community priorities into public decisions in shared waters. To be meaningful, public participation processes should be designed to encourage involvement among the full range of stakeholder interests, including underserved communities such as women, the poor, and indigenous communities that are or would be directly or indirectly impacted by a hydraulic project. Communities historically excluded from public processes must be brought into the public-participation process, and direct investments in their capacity development should be made to ensure their meaningful participation. These communities are key stakeholders in water diplomacy and must be engaged in meaningful ways at every possible opportunity—despite being historically unheard in the public discourse of water negotiation.

Water negotiations typically involve decision makers who are political leaders focused on preserving sovereignty and maintaining state security despite the fact that they are often unprepared to think about improving the overall efficiency of water use.

Competition over shared waters should also warrant strong interest from foreign policymakers. Foreign policy can help improve transboundary water governance, and transboundary water governance can provide policymakers a toehold for making progress on crucial foreign policy interests. Thus, encouraging greater cooperation over transboundary waters offers significant prospects for the resolution of political conflicts and greater regional integration. Transboundary waters constitute a promising entry point for diplomats aiming for high peace dividends. Foreign policymakers can and should do more to realise these dividends.

Water diplomacy can involve actors other than traditional state actors, including civil society organizations or academic networks.

And yes, career water diplomats are also needed, but this role has historically been in the context of transboundary waters. We envision that water diplomats will be a new type of water professionals who will be competent in multiple water-related academic and applied fields. A key challenge is how to develop the competencies as an interdisciplinary skill set for youths entering a water diplomacy program designed for them with diverse undergraduate preparation in natural sciences, social sciences, and engineering. While we see this as a particularly needed skill set, it is additionally imperative that youths exiting the water diplomacy program have an array of appropriate career opportunities open to them, as "water diplomat" is not yet a recognizable career option. Therefore, each youth has to be an expert in his or her discipline as well as having the skills to work effectively across disciplines.

Government-established transboundary river basin organizations can act as connectors and active peacemakers by facilitating inclusive dialogues and participatory decision-making. River basin organizations have established mechanisms for multi-stakeholder engagement, giving voice to young people, women and concerned stakeholders, some of whom have been expressing support for the appointment of a special envoy on water, stating that promoting water should be one of the United Nations priorities for the years to come and a strong voice that can really mobilize the international community is needed.

If applied appropriately, water diplomacy will bring more actors to the table, increasing knowledge, and facilitating dialogue. It is hoped that by working together—on the basis of the two United Nations water conventions—and by developing shared partnerships with civil society and the private sector, the international community can overcome water-related challenges and build a more prosperous future.

Access to More Working Knowledge for Better Communication

1. Read the facts listed below and discuss in pairs or at a mock negotiating session about how we can mobilize action in order to promote water diplomacy for the sake of peace and sustainable development.

—UN Water has emerged as an interagency mechanism that aims to coordinate other UN agencies working on water issues.

—Given its limited resources and standing as a mechanism, however, it is questionable whether UN Water can play a role in coordinating proactive hydro-diplomacy.

—China directly provided flood-season hydrological data of the Lancang River to the Mekong countries and offered timely updates on upstream flow changes. Despite its own difficulties, China increased outbound water flow in the drought season to help downstream countries mitigate droughts.

—Water-related problems are complex not only because they involve various stakeholders (e. g. , farmers, industrial users, urban developers, environmental activists) who are competing for a limited resource, but also because water problems cross multiple boundaries (e. g. , natural, societal and political), scales (spatial, temporal, jurisdictional and institutional) and levels (local, regional and global) that are difficult to incorporate into a traditional modeling approach.

—To support the negotiation process, the committee works with a neutral facilitator and invites other development partners from the basin to participate in the negotiation process.

—Selected water diplomacy actors include career diplomats.

—However, shaping the agreements and tools for transboundary waters requires both technical and diplomatic expertise, necessitating a hydro-diplomacy approach that includes technical experts with skills in negotiation and mediation as well as policymakers and diplomats who have knowledge of water management and the development of water infrastructure.

—Water diplomats are the water professionals of tomorrow.

—In acknowledging the principles of others, a water diplomat does not have to sacrifice his or her own principles.

—The main water diplomacy actors typically consist of riparian country governments and related intergovernmental organizations such as river basin commissions, often coupled with other actors across multiple diplomacy tracks. Several documents also imply the engagement of so-called third-party actors such as external countries, networks or organizations in the water diplomacy processes. These actors can build trust and facilitate water diplomacy related processes through joint capacity-and-knowledge-building, network activities, facilitation and mediation.

—Broadening the group of current decision makers to include youth perspectives is important—dealing with science can be costly and expensive, but training water experts, water diplomats, especially in the developing world is important.

—Today's youth are not only direct beneficiaries of the existing policies concerning water governance, but they are also the ones that need to work to ensure access to clean water in the coming decades, if not centuries.

—Their objective is to educate youth with basic understanding of technical and hydrological knowledge to become "water diplomats", who will be the ones in the future to work with various stakeholders to promote international cooperation in the field of water resources.

—Many small developing riparian countries suffer from a lack of competent hydro-diplomats who can address climate change issues while carrying out negotiations over shared water resources in the context of existing power asymmetries.

2. Find out who else have been involved in water diplomacy besides the following list of people or institutions and talk about what they have done in intercultural communication with interested parties on global water issues.

United Nations Environment Programme (UNEP) 联合国环境规划署

United Nations Development Programme(UNDP) 联合国开发计划署

International Network of Basin Organizations(INBO) 国际流域组织网络

United Nations Water(UN Water) 联合国水机制

United Nations Educational, Scientific and Cultural Organization-Intergovernmental Hydrological Programme(UNESCO-IHP) 联合国教科文组织政府间水文计划

United Nations Economic Commission for Europe(UNECE) 联合国欧洲经济委员会

monitoring agreements through a joint authority 通过联合管理机构进行监督的协议

trained water professionals 训练有素的水利专业人员

water and sanitation sector professionals 水利部门、卫生部门专业人员

multilateral donors 多边捐助者；多边赠予国；多边赠予机构

bilateral donors 双边捐助者；双边赠予国；双边赠予机构

international donor agencies 国际捐助机构

the International Association for Water Law(AIDA) 国际水法协会

UNESCO World Water Assessment Programme(WWAP) Secretariat 联合国教科文组织世界水资源评估计划秘书处

UN-Water family 联合国水机制大家庭

UN-Water summit on groundwater 联合国水机制地下水峰会

Stockholm Water Institute(SIWI) 斯德哥尔摩水利研究所

Stockholm Junior Water Prize 斯德哥尔摩青少年水奖

Young Water Diplomats(YWD) Programme 青年水外交官计划

non-governmental organizations(NGOs) 非政府组织

a diverse array of peace/conflict resolution organizations(P/CROs) 各种和平/冲突解决组织

the elite-level diplomats 精英外交家

Interstate Commission for Water Coordination in Central Asia (ICWC) 中亚国家间水资源协调委员会

Shared Waters Partnership 共享水域伙伴关系

Shared Waters Cooperation Facility 共享水域合作机制

International Centre for Water Cooperation(ICWC) 国际水合作中心

UNDP Water Governance Facility(WGF) 联合国开发计划署水治理机制

3. Read the case below, thinking about how to learn to be empathetic and use water diplomacy to complement governmental diplomacy.

In Yemen, the water of the Malaka Dam was primarily used by three neighbouring villages for irrigation and livestock, and was a subject of conflict for decades. In an attempt to halt the conflict, a tribal decree was put in place forbidding all use of the dam water. After that, a water user association (WUA) managed by women in the community, Al Malaka, took the lead in dispute resolution and peace negotiation surrounding the dam water usage. WUA members, with support from the Food and Agriculture Organization of the United Nations(FAO), were able to negotiate the implementation of a piping system that would use gravity flow to send the Malaka Dam water to several groundwater wells in the area. This solution was innovative and effective in that it eliminated the need for direct use of the dam water, while it decreased evaporation and rejuvenated well water resources. The water has since been used peacefully for livestock and irrigation in the surrounding areas. This example highlights the need for community involvement and the inclusion of women in matters of water diplomacy in the Arab region. The FAO has continued to build on the success of the Malaka Dam story by implementing the project "Water for Peace in Yemen: Strengthening the Role of Women in Water Conflict Resolution". Funded by the United Nations Secretary-General's Peacebuilding Fund, this project was implemented by the FAO, the International Organization for Migration(IOM), the Hadramout Ministry of Water and Environment, and the Women Water Users Group(WWUG). While ongoing tensions necessitated a change in the original target area of the project from the Hudaydah governorate to the Wadi Hadramout in the Hadramout governorate, the project has since shown several markers of success. Similar to the Malaka Dam project, women were involved as peace mediators to resolve disputes surrounding water use. One of the most significant outcomes of the project was that it allowed families to return to farming lands that had previously been abandoned due to a lack of irrigation. Female farmers also reported significant changes in water availability and conflict prevention between neighbouring communities.

Further Recommended Reading

[1] Jarvis, W. Todd. *Contesting hidden waters: conflict resolution for groundwater aquifers*. New York: Routledge, 2014.

[2] United Nations Development Programme. *Capacity development: a UNDP primer*. New York: UNDP, 2015.

[3] Shafiqul Islam, and Kevin M. Smith. *Interdisciplinary collaboration for water diplomacy: a principled and pragmatic approach*. New York: Routledge, 2020.

[4] Islam Shafiqul, and Susskind, Lawrence E. *Water diplomacy: a negotiated approach to managing complex water networks*. 1st ed. New York: Routledge, 2013.

Part Ⅱ

Cases of Water Diplomacy between Riparian States

A general problem with case studies is that they capture a history up to a point in time, and then they are published and put aside. Cases are usually limited in terms of the perspectives or framing for the case problem. A traditional case study represents what might only be a small piece of a larger puzzle that requires additional context, new perspectives, and updates over time to truly understand.

Part Ⅱ is an opportunity to make case studies of transboundary water problems and their solution more like living documents, with the ability to add and update. Its goal is to create a place where cases could represent multiple viewpoints in intercultural communication activities like water negotiations, cooperation and exchanges and be edited over time to add greater context. Rather than attempting to only include consensual views or those informed by our own limited researches and biases, we want to make space for unusual or opposing takes on the same problem.

Transboundary water events often involve complex negotiations, technological challenges, and multifaceted governance structures. While institutional frameworks and geopolitical considerations dominate the discourse, individual actors—government representatives, scientists, activists, and community leaders—play pivotal roles in shaping outcomes. Individuals bring unique skills, relationships, and personal values to bear on processes that determine whether water disputes escalate into conflicts or evolve into opportunities for cooperation. Examining cases can illuminate the critical role of individuals in transboundary water management.

Keep in mind UN Secretary-General António Guterres's observation that "Water is catalyst for cooperation, not conflict", so that we might make water cooperation our common language and concern.

Meanwhile, keep in mind the following questions while studying this part: How could one case culturally relate to water conflicts? Will they become future problems? What could one case do toward informing individuals with interest or stakes in the other cases? What key issues or questions would forge a meaningful linkage between the case and other complex water problems?

Unit 4 Transboundary Water Disputes and Conflicts

Background Study for Successful Intercultural Communication

Read authentic Chinese and English topical texts in advance before making a glossary for their associated terminology and jargon as follows.

transboundary water disputes 跨界水资源争议

water conflict chronology 水冲突年表

shared water resources 共享的水资源

Tigris-Euphrates Basin 底格里斯和幼发拉底河流域

political stability and economic development 政治稳定与经济发展

watercourses 水道

Keban Dam 凯班坝(凯班水库)

Tabqa Dam 塔卜卡大坝(又名 Euphrates Dam 幼发拉底水坝)

Middle East 中东

Shatt al-Arab waterway 沙特阿拉伯水道

Basra, Iraq 伊拉克巴士拉(伊拉克巴士拉省省会，是伊拉克第一大港及第二大城)

Persian Gulf 波斯湾

river basins 河流流域

marshlands, reed beds, and islands 沼泽地、芦苇荡和岛屿

cornerstone of regional stability 区域稳定的基石

transboundary aquifers 跨界含水层

increasing risks of water conflict 水冲突风险增加

geopolitical tensions 地缘政治紧张

Indus Waters Treaty《印度河水域条约》

hydropower projects 水电项目
upstream developments 上游开发
environmental degradation 环境恶化
Nile Basin governance 尼罗河流域治理
climate variability 气候变化
water scarcity 水资源短缺
Jordan River Basin 约旦河流域
water equity issues 水资源公平问题
frameworks for cooperation 合作框架
adaptive governance 适应性治理
groundwater dependency 对地下水的依赖
socioeconomic conditions 社会经济条件
water management frameworks 水管理框架
interstate conflicts 国家间冲突
intrastate tensions 国内紧张局势
water rights violations 水权侵犯

Text A
Water Disputes and Conflicts in the Nile River Basin

Pre-reading Questions

1. Have you personally experienced any disputes or conflicts involving water access or usage? Or have you ever come across any instances of water-related violence mentioned in your daily life, the media, or books?

2. When and where do you believe water disputes and conflicts are most likely to arise, and what factors have influenced your thinking?

3. What do you think are some cultural beliefs or values that are related to water? How might those perspectives influence the ways communities perceive, manage, and use water resources?

4. In cultures where large families are valued as symbols of wealth and prosperity, high fertility rates often lead to increased water demand and straining

existing resources. However, urbanization and changing social norms have led to declining interest in large families in some regions. How do varying cultural attitudes towards population growth and resource use affect water management in different regions?

The Nile River Basin, a critical source of freshwater for millions of people across Northeast Africa, has long been a focal point for water disputes and conflicts. These conflicts are primarily driven by water scarcity, the burgeoning demand resulting from growing populations, and the increasingly severe impacts of climate change. The Nile River Basin is a dynamic system that represents a potential source of conflict given its finite water resources, growing population, challenges with food security, and reliance on hydroelectric power as a major energy source.

Conflicts over Nile water allocation have deep roots in the colonial history of the region. The 1929 Nile Water Agreement, signed during the colonial period, allocated the majority of the Nile's water to Egypt and Sudan, granting Egypt the right to use 48 billion cubic meters of water annually while Sudan was allocated only 4 billion cubic meters. The agreement also gave Egypt veto power over any water projects in upstream countries, severely limiting the water rights of other Nile Basin countries.

In 1959, an updated agreement further solidified Egypt's dominance over Nile water resources, allowing it to use 55.5 billion cubic meters of water annually while Sudan was allocated only 18.5 billion cubic meters. The colonial-era agreements favored Egypt and Sudan, leaving upstream countries like Ethiopia, Uganda, and Kenya with minimal access to Nile water resources. These countries argue that the agreements are discriminatory and hinder their agricultural and development ambitions. They also argue that the treaties are invalid because they were imposed by colonial powers without their consent.

In 2011, Ethiopia initiated the construction of the Grand Ethiopian Renaissance Dam (GERD) as its growing domestic capabilities and investments from the Ethiopian diaspora enabled the project. This shift challenged Egypt's historical dominance over the Nile, marking a turning point in regional water politics.

The Nile Basin Initiative (NBI), established in 1999, aims to promote sustainable development and regional cooperation in the management of the Nile's water resources. However, the negotiation of the Cooperative Framework Agreement (CFA) has led to an impasse, with Egypt and Sudan withdrawing from NBI projects in 2010. Sudan returned to the NBI in 2012, recognizing the importance of basin-wide cooperation, especially after the GERD project was launched.

Egypt's withdrawl is not without a deeper national concern. The country is extremely worried about the GERD, as it depends on the Nile River for 97% of its water supply, with nearly 100 million people relying on the river for drinking water, agriculture, and industry. The Nile is considered Egypt's lifeline, deeply embedded in its cultural and economic identity. Any reduction in water flow could have catastrophic consequences for its population and economy. The country fears that the GERD will significantly reduce water flow to its territory. Ethiopia's unilateral filling of the dam's reservoir (which began in 2020) and its plans for future fillings could exacerbate Egypt's water scarcity crisis. Studies suggest that even a 1 billion cubic meter reduction in Nile water could eliminate over 1 million jobs and cause $1.8 billion in annual economic losses. Its Prime Minister Mostafa Madbouly emphasized that the dispute is "related to the survival of the Egyptian people for generations to come". Therefore, Egypt has consistently opposed the dam's construction and filling, bringing the issue to the United Nations Security Council in 2024.

Between 2011 to 2015, Sudan initially aligned with Egypt in opposing the dam's construction, fearing reduced water flow and impacts on its irrigation systems. In 2015, Ethiopia, Sudan, and Egypt reached a landmark agreement known as the "Declaration of Principles" (DoP), which provided a framework for continuing the GERD construction and cooperation based on the principles of equitable utilization and no significant harm. Since that trilateral agreement, Sudan has shifted its stance to support the GERD, recognizing its potential benefits for hydropower and flood management. In 2023, Sudan's de facto leader Abdel Fattah al-Burhan stated that Sudan was "aligned and in agreement" with Ethiopia on the GERD issue. However, Sudan remains concerned about the lack of regulations governing the dam's operation and fears that unregulated management could lead to flooding, particularly in areas already vulnerable to seasonal water level fluctuations.

The GERD, with an expected capacity of 5,150 MW, is estimated to double Ethiopia's electricity generation and improve access to energy within the country. This surplus energy could be traded with neighboring countries, providing them with an alternative energy source.

Energy trade between Ethiopia, Sudan, and Egypt presents a promising avenue for cooperation due to its potential to address water scarcity, boost hydropower generation, reduce greenhouse gas emissions, and foster regional economic integration. By trading electricity, those countries can reduce reliance on water-intensive energy sources (e. g., thermal plants) and optimize water use. For example, Ethiopia's hydropower generation could help Egypt meet its energy needs while preserving Nile water for agriculture. While Egypt has an electricity surplus (25% in 2022), engaging in energy trade could support its goal of becoming a regional energy hub and reduce tensions with Ethiopia. Sudan also stands to benefit from affordable electricity imports, reducing reliance on thermal plants and expanding irrigation capacity. Improved energy supply could boost its agricultural productivity and GDP by $27~29 billion by 2060.

International donors and organizations, such as the European Union and Germany, have already provided financial and technical assistance to initiatives like the NBI. Sustained international engagement is crucial for building institutional capacity within the Nile Basin states. This includes training in water management, data collection and analysis, and conflict resolution. International support can also facilitate dialogue among the riparian states, helping to bridge the gaps between different interests. Moreover, it can ensure the adoption of sustainable water management practices that are in line with best global standards.

The Nile River Basin is a region of vital importance yet one that is fraught with volatility. Water scarcity, population growth, and climate change have combined to create significant challenges for the sustainable management of the Nile's resources. While historical treaties and large-scale projects like the GERD have, in some ways, contributed to the existing disputes, efforts such as the NBI and the concept of cooperative adaptive management offer hope for fostering regional collaboration.

To ensure a more sustainable future, it is essential to address the environmental and socio-cultural impacts of water scarcity. This will require a shared sense of responsibility among the riparian states, the establishment of effective benefit-sharing mechanisms, and continued international support. Only through these collective efforts can water security and stability be achieved for the diverse communities that call the Nile Basin home.

Access to More Working Knowledge for Better Communication

Read the following facts from UN Water Reports and further understand factors that may either aggravate water disputes and conflicts or lead to cooperation among riparian states and countries.

—Water is at the center of economic and social development; it is vital to maintain health, grow food, manage the environment, and create jobs. Despite the importance of water, globally, 2 billion people lack access to safely managed drinking water and 3.6 billion people lack access to safely managed sanitation.

—The sum of surface runoff and groundwater recharge (total runoff) over a certain period represents the renewable water resources, equal to precipitation minus evapotranspiration (effective precipitation). It is the amount of water we can use for various human needs, such as drinking, irrigation, industry, and tourism and store in surface reservoirs or in groundwater aquifer systems by artificial recharge, without compromising the natural water resources reserves of the area (sustainable use of water resources).

—Hydrometeorological and hydrological data collected at a national scale need to be harmonized if not supported by a common monitoring system.

—Hydrological upstream-downstream asymmetry and different hydrogeological interactions between the riparian countries generate the main problem of sharing the amount of water flowing from one country to another.

—Water quality issues may induce severe problems between riparian countries, especially when upstream countries export polluted water to downstream countries.

—Hydrological extremes, such as floods and droughts originated in one riparian country, may cause harm to another located downstream.

—Different national water laws and related policies may create challenges for managing water use and protecting water quality in transboundary water basins. For example, upstream countries may neglect wastewater treatment, which may cause water pollution problems to downstream countries.

—Different socioeconomic conditions between riparian countries mean different needs for water in transboundary river catchments and different allocation between economic sectors, such as agriculture, tourism, industry, and environmental protection of shared water. For example, a poor country located upstream may need to use more water, hence constructing dams that reduce water availability downstream, which may be detrimental to a riparian country located downstream.

—Conflicting political objectives and different political priorities between countries may lead to a lack of trust and a failure to develop mutual interests.

—Application of international law and policy in transboundary water basins depends on political considerations for ratifying international, regional or bilateral conventions and agreements.

Development of Intercultural Communication Skills

1. In some European folk traditions, water is associated with spring festivals. In Italy, the festival of San Benedetto in Norcia features the blessing of water sources. The water is believed to have purifying and curative properties, and people collect it for good luck. What do you know about the different views of water that various cultures have in their daily lives, traditions, and spiritual practices?

2. The Turks favor settling in areas near water, as water is considered a sanctuary offering protection from harm and illness, a means of purification and cleansing, and, ultimately, a source of healing. What are the widely accepted beliefs or values regarding water as a shared resource in your culture?

3. Islam is the predominant religion in Turkey. Water is valued as a symbol of cleanliness and life in the Islamic faith. Water supply, necessary to human

communities, can ensure health and hygiene conditions and support various creative activities. How can an understanding of water's significance in Turkish culture help avoid misunderstandings in cross-cultural settings?

4. What cultural considerations should be kept in mind when discussing water-related issues (such as sanitation or hygiene) in intercultural communication scenarios?

5. The high-context nature of Arabic culture indicates that non-verbal cues often carry as much weight as the spoken word. What should potential water diplomats pay special attention to when dealing with counterparts from riparian states or communities of Islamic faith?

6. Given Turkey's use of water for both civic and cultural purposes, how can effective resource management and distribution contribute to fostering intercultural communication on shared challenges such as access to clean water?

7. In November 2023, the Turkish Agriculture Minister warned that the country is on the verge of a severe water crisis, with projections suggesting that by 2030, Turkey's water availability will drop below 1,000 cubic meters per capita. As a future water professional, what measures do you propose to mitigate the impact of this situation?

Text B

The Indus River Basin Development from Disputes to Diplomacy

Pre-reading Questions

1. How do economic interests related to water resources contribute to tensions or collaboration among riparians? What cultural values might influence those interests?

2. What innovative approaches do you think could lead to greater cooperation over shared water resources in the future? How can cultural understanding facilitate these efforts?

3. What successful stories of water cooperation can you think of? What inter-cultural communication or diplomatic lessons can be learned from these examples

that might be applicable in other contexts?

4. How does media representation of water conflicts and cooperation influence public perceptions across cultures? What role does media play in shaping diplomatic talks?

The Indus River Basin serves as a prominent example of a transboundary water system with a history marked by both conflict and cooperation. The roots of the Indus Water Basin conflict can be traced back to the partition of British India in 1947, which left India and Pakistan to navigate the division of vital resources, including water. Central to the dispute is the territorial status of Kashmir, claimed by both nations and divided by the Line of Control (LoC). The flow of the Indus River's tributaries significantly influenced the demarcation of this boundary, underscoring the region's strategic importance.

After their independence from British control, water resources became a key source of tension between India and Pakistan. The partition positioned India upstream, granting it control over barrages that regulate water flow into Pakistan. This upstream-downstream dynamic heightened concerns in Pakistan, where about 90% of food production and 65% of the workforce depend on agriculture and livestock reliant on the Indus River. As demands for water for irrigation and power grew, tensions between India and Pakistan intensified, particularly over dam projects in Indian-controlled areas.

To maintain stability during the transitional period, both nations signed a short-term Standstill Agreement in 1947. However, after this agreement expired, India began diverting water from canals flowing into Pakistan on April 1, 1948. This led to the Inter-Dominion Accord of May 4, 1948, which mandated that India supply water to Pakistan's portion of the Indus River Basin in exchange for annual payments. Although soon interrupted by stalled negotiations, the Accord represented an initial attempt at cooperation and laid the foundation for the Indus Waters Treaty.

Bilateral cooperation on transboundary waters also gained momentum through external mediation. In 1951, David Lilienthal, former head of the Tennessee Valley Authority and the U. S. Atomic Energy Commission, proposed joint development and management of the Indus River system with the help of the World

Bank. Despite political barriers, the idea gained support from Eugene Black, then president of the World Bank. Engineers from India and Pakistan formed a technical group with support from the World Bank, although progress was stymied by political challenges. In 1954, the World Bank submitted a proposal to resolve the stalemate.

After six years of intense negotiations, the Indus Waters Treaty (IWT) was signed in September 1960 by Jawaharlal Nehru, the Indian Prime Minister, Muhammad Ayub Khan, the Pakistani President, and Sir William Iliff from the World Bank. In spite of its compromising spirit, the Treaty significantly alleviated the tension and paved the way for practical water management projects. The Treaty allocated control of three rivers—Indus, Jhelum, and Chenab—to Pakistan and three rivers—Ravi, Beas, and Sutlej—to India. Furthermore, it established the Permanent Indus Commission, a framework for resolving disputes, with the World Bank acting as a mediator in case conflicts should persist.

The successful conclusion of the Indus Waters Treaty provided a lasting resolution to the 13-year-long water dispute. It created a legal framework for managing and using water resources in the Indus River Basin, transforming the dynamics of India-Pakistan water interactions. Its provisions for data sharing and joint commissions fostered cooperation and transparency, playing an essential role in maintaining peace and stability between the two nations.

However, despite periods of political hostility, the Indus Waters Treaty has largely facilitated consistent water-sharing arrangements. Still, India's upstream water infrastructure projects have occasionally reignited tensions. For example, the Baglihar Dam on the Chenab River became a point of contention, with Pakistan accusing India of violating the Treaty by potentially restricting water flow. Arbitration permitted the project to continue, albeit with design adjustments, underscoring the Treaty's role in resolving disputes. Similarly, the Wular Barrage or Tulbul Navigation Project on the Jhelum River, proposed by India to improve navigation, was met with objections from Pakistan, who feared reduced water flow. These disputes highlighted the need for modernizing the Treaty to address ongoing and future challenges.

Transboundary waters present both risks and opportunities. While water allocation disputes and infrastructure conflicts are common, the Indus River Basin illustrates that cooperation can emerge from conflicts through trust-building, institutional frameworks, and shared benefits. By drawing lessons from successful cooperation models, the chances of turning water conflicts into opportunities for sustainable development are significantly improved.

Access to More Working Knowledge for Better Communication

1. Read the facts listed below and discuss in pairs or at a mock negotiating session about how we can mobilize action in order to promote awareness in the importance of transboundary water cooperation.

—There are 263 transboundary river basins and approximately 300 transboundary aquifers. (UNECE/UNESCO 2015)

—145 states have territory within transboundary lake or river basins, and 30 countries lie entirely within them. (UNECE/UNESCO 2015)

—Since 1948, history shows only 37 incidents of acute conflict over water, while during the same period, approximately 295 international water agreements were negotiated and signed. (UNECE/UNESCO 2015)

—Around two-thirds of the world's transboundary rivers do not have a cooperative management framework. (SIWI)

—Promoting transboundary water cooperation aligns with global goals such as the Sustainable Development Goals (SDGs), particularly Goal 6 (Clean Water and Sanitation) and Goal 16 (Peace, Justice, and Strong Institutions).

—Effective transboundary water cooperation can lead to improved governance structures and institutional frameworks, ensuring that water management is transparent, participatory, and accountable.

—Raising awareness about transboundary water cooperation can help prevent crises before they escalate into conflicts, allowing for proactive measures and negotiations rather than reactive responses. Transboundary cooperation facilitates the exchange of technology and knowledge, enabling countries to implement best practices in water management, such as efficient irrigation techniques and pollution control.

—Successful cooperation over water resources can promote social stability and peacebuilding by fostering dialogue and mutual trust among nations. Water diplomacy can serve as a platform for broader political cooperation.

—Collaborative management of transboundary waters is crucial for adapting to climate change impacts, such as altered precipitation patterns and extreme weather events. Countries can share data and resources to enhance resilience.

—There are over 3,600 international treaties and agreements related to water, demonstrating the recognized need for cooperation. Promoting awareness can facilitate the development of new agreements and strengthen existing ones.

—Approximately 40% of the world's population lives in river and lake basins that cross national borders. Effective management of these transboundary water resources requires cooperation to avoid disputes and ensure sustainable use.

2. Find out more about the following organizations and institutions dedicated to promoting transboundary water cooperation and sustainable management of shared water resources and talk about what they have done in intercultural communication with interested parties on global water issues.

Global Water Partnership(GWP) 全球水合作伙伴关系

International Water Management Institute(IWMI) 国际水管理研究所

International Hydropower Association(IHA) 国际水电协会

International Commission on Irrigation and Drainage(ICID) 国际灌溉与排水委员会

International Association for Water Law(AIDA) 国际水法协会

International Union for Conservation of Nature(IUCN) 国际自然保护联盟

Water Resources Institute(WRI) 水资源研究所

Transboundary Water Resources Management Institute(TWRMI) 跨界水资源管理研究所

Water Integrity Network(WIN) 水透明网络

Global Environment Facility(GEF) 全球环境基金

Asian Development Bank(ADB) 亚洲开发银行

Food and Agriculture Organization(FAO) 食品和农业组织;粮农组织

Global Water Security Coalition 全球水安全联盟

Water and Sanitation Program(WSP) 水与卫生计划

Danish Institute for Human Rights(DIHR) 丹麦人权研究所

Eurasian Water Network 欧亚水网络

Regional Center for Development Cooperation（RCDC）区域发展合作中心

Green Cross International 国际绿十字会组织

Rivers of the World（ROW）世界河流

Water Governance Facility（WGF）水治理设施

International Institute for Sustainable Development（IISD）国际可持续发展研究所

Water Policy Program（WPP）水政策项目

Center for Water Security and Cooperation 水安全与合作中心

Mediterranean Information Office for Environment, Culture and Sustainable Development（MIO-ECSDE）地中海环境、文化和可持续发展信息办公室

Water Supply and Sanitation Collaborative Council（WSSCC）水供给与卫生合作理事会

Water Governance and Sustainability Network 水治理与可持续发展网络

International Center for Water Resources Management（ICWRM）国际水资源管理中心

Council of Europe 欧洲理事会

African Water Association（AWA）非洲水协会

Further Recommended Reading

[1] Allan, John A. *Water wars: conflict and cooperation along international borders*. New York: MIT Press, 2001.

[2] King, Marcus Dubois. *Water and conflict in the Middle East*. Oxford: Oxford University Press, 2020. https://doi.org/10.1093/oso/9780197552636.001.0001.

[3] Pacific Institute. *Water conflict*. Oakland: Pacific Institute, 2024. https://www.worldwater.org/water-conflict/.

Unit 5 Transboundary Water Resources Management

Background Study for Successful Intercultural Communication

Read authentic Chinese and English topical texts in advance before making a glossary for their associated terminology and jargon as follows.

Murray-Darling Basin 墨累-达令盆地

Murray River 墨累河

Darling River 达令河

Sustainable Diversion Limit (SDL) 可持续分流限额

Murray-Darling Basin Plan 墨累-达令盆地计划

Murray-Darling Basin Authority (MDBA) 墨累-达令盆地管理局

Sustainable Diversion Limit Adjustment Mechanism (SDLAM) 可持续分流限额调整机制

catchment 集水区

wetlands 湿地

endangered species 濒危物种

First Nations 原住民

irrigation 灌溉

Integrated Water Resources Management (IWRM) 水资源综合管理

interconnected rivers and lakes 互联互通的河流和湖泊

infrastructure improvements 基础设施改进

equitable distribution 公平分配

stricter environmental regulations 更严格的环境法规

improved wastewater treatment facilities 改善的废水处理设施

data sharing 数据共享

joint problem-solving 联合解决问题

environmental water reserves 环境水储备
stakeholder engagement 利益相关者参与
Australian Bureau of Statistics (ABS) 澳大利亚统计局
water resources management 水资源管理
flood and drought response 洪水和干旱应对
effective governance 有效治理
transboundary basin 跨界流域
water-related disasters 水相关灾害
emergency situations 紧急情况
ecosystem degradation 生态系统退化
groundwater levels 地下水位
desertification 沙漠化
aquatic ecosystems 水生生态系统
Chu-Talas Rivers Commission 楚河-塔拉斯河水利委员会
water management facilities 水管理设施
vulnerability assessment 脆弱性评估
runoff 径流
glacier melt 冰川融化
Turan lowland 图兰低地

Text A
Water Resources Management in the Murray-Darling Basin

Pre-reading Questions

1. Can you offer examples of how communities which embrace the philosophy that "the water that bears the boat can sink it" would approach the management and conservation of shared water resources?

2. What diplomatic strategies have been successful in promoting cooperation over shared water resources? How can those strategies be tailored to align with the unique cultural contexts of different regions or riparians?

3. What lessons can be learned from global practices regarding conflict

resolution in transboundary water disputes? How can these lessons inform future diplomatic negotiations?

4. What roles do international organizations (e. g. , UN, World Bank) play in promoting effective water resources management? How can those organizations employ the cultural sensitivity in their approaches?

The Murray-Darling Basin, situated in southeastern Australia, contains 22 major catchments, with the Murray and Darling Rivers as the two primary waterways. Originating in southern Queensland at the confluence of the Culgoa and Barwon Rivers, the Darling flows into the Murray along the New South Wales and Victoria border. Then the Murray continues its journey to the sea, southeast of Adelaide.

The basin, culturally, ecologically, and economically significant, is home to over 2. 4 million people, including individuals from more than 50 First Nations. Its wetlands, spanning 16 internationally recognized sites, support an incredible diversity of life, covering 35 endangered species and 120 species of waterbirds. Around 55% of Australia's total water use is concentrated in this region, primarily for agricultural purposes using surface water sources.

Conflicts over water allocation in the basin have existed since colonial times, often complicated by arbitrary state borders drawn by distant colonial administrators. In recent decades, states like South Australia have demanded larger water shares, prompting opposition from upstream states like New South Wales and Victoria.

The Basin has also served as a testing ground for innovative water management and policy initiatives, which has shaped Australia's national water reform programs, particularly since the 1990s.

The Murray-Darling Basin Authority (MDBA) was established in 2007 under the Water Act 2007 as an independent body to implement the Murray-Darling Basin Plan. Its purpose is to plan the Basin's water resources as a whole and to work with Basin states, industries and communities.

A key responsibility of the MDBA is to manage environmental water holdings, helping protect and restore the health of rivers, wetlands, and floodplains. In line with the Murray-Darling Basin Plan, the MDBA regulates water extraction and usage, enforcing compliance with established rules and caps. To achieve effective water resource management, the MDBA collaborates closely with state governments, water users, indigenous communities, and other stakeholders.

Through the collection and analysis of data on water availability, use, and environmental health, the MDBA formulates its planning and management decisions. Computer modeling allows scientists to analyze various scenarios and forecast potential impacts on water flow patterns, water quality, and ecosystem health.

The MDBA also engages extensively with communities to understand their needs and concerns, promoting transparency and inclusivity in decision-making. Regional consultations and informational events are made available on the state government websites to promote public engagement activities. Moreover, it has Regional Engagement Officers based in different basin regions to share responsibility in listening to and advocating for the needs and priorities of the communities they represent, such as focusing on sustainable outcomes and the associated environmental, social, and economic issues. Additionally, recognizing the challenges posed by climate change, the MDBA is developing strategies to enhance the resilience of the basin's water management systems.

The MDBA also works as a mediator of disputes over water allocation and use among states and user groups, aiming to balance competing demands fairly. To uphold its commitment to transparency, it provides regular updates and reports on the status of water and the progress of the Basin Plan. This comprehensive approach ensures that the Murray-Darling Basin is managed to benefit current and future generations.

In conclusion, the management of water resources in the Murray-Darling Basin faces both substantial challenges and promising opportunities. Balancing agricultural, environmental, and community demands requires strategic innovations and collective actions. The Murray-Darling Basin Plan, along with the work of the Murray-Darling Basin Authority, has laid the groundwork for integrated water management. However, challenges such as climate change, excessive water extraction, and ongoing interstate conflicts highlight the need for adaptable solutions and stronger cooperation. By prioritizing participatory decision-making, advancing scientific research, and embracing sustainable practices, the MDBA has the potential to set a global example in managing shared water resources while preserving its vital ecosystems for future generations.

Access to More Working Knowledge for Better Communication

1. Read the facts listed below and discuss in pairs or at a mock negotiating session about how can water diplomats help to improve transboundary river management with intercultural communicative skills.

—Transboundary management often focuses on resources that span borders, such as rivers, lakes, forests, and wildlife. Effective management is essential to ensure equitable use and conservation of these resources.

—It involves cooperation between neighboring countries or regions, requiring diplomatic engagement, joint decision-making, and collaborative action to address shared challenges.

—Transboundary management employs an integrated approach that considers ecological, social, economic, and political factors. This holistic perspective is essential for sustainable resource use and environmental protection.

—By establishing frameworks for communication and negotiation, transboundary management can help prevent and resolve conflicts over shared resources, which might arise due to competing interests.

—Effective transboundary management often relies on treaties, agreements, and institutions that provide guidelines for cooperation and set the rules for resource management.

—Rivers like the Ganges-Brahmaputra, Nile, and Danube require transboundary management to address issues such as water quality, allocation, and flood control.

—Protected areas that span borders, such as transboundary parks, necessitate joint management efforts to conserve biodiversity and facilitate wildlife migration.

—Air pollution that crosses borders (e. g., from industrial emissions) requires coordinated policies and regulations to mitigate its impact on health and the environment.

—Transboundary regions must collaborate to develop strategies for climate change adaptation, such as shared infrastructure for disaster risk reduction and joint research initiatives.

—Coordinated management promotes the sustainable use of shared resources, ensuring that future generations can also benefit from them.

—Effective cooperation can help build trust among countries, reducing the likelihood of conflicts and enhancing regional security.

—Joint efforts can lead to better environmental protection and restoration, benefiting ecosystems and local communities.

—Collaborative management can create economic opportunities through joint ventures, tourism, and sustainable practices that enhance regional development.

2. Examine the following common myths about transboundary river management. Use well-reasoned arguments to dispel these misconceptions and propose actionable strategies to promote effective transboundary water management practices among all stakeholders involved.

Myth 1: Transboundary issues are exclusively about negative things such as conflicts, hatred, and even wars.

Myth 2: A single country can effectively manage shared resources without any cooperation from its reparians.

Myth 3: Managing transboundary waters has nothing to do with the concept of water sovereignty.

Myth 4: Ordinary men and women with hydraulic literacy have no say in transboundary resources management and conservation.

Myth 5: All shared resources including water hold equal value for neighboring countries.

Myth 6: Transboundary resource management only pertains to water.

Myth 7: Solution to transboundary water issues must always be a top-down process.

Myth 8: Transboundary cooperation is doomed to be ineffective, impractical and eventually a failure.

Myth 9: Climate change and environmental challenges are irrelevant to transboundary water issues, including floods, droughts, and pollution.

Myth 10: Water diplomacy from an intercultural communicative perspective is nothing but a recent idea.

3. Discover more about the institutional framework that is established to facilitate the collaboration and shared governance in managing transboundary waters, you may start with the following treaties and agreements.

Kyrgyzstan-Uzbekistan Water Agreement (1998)

Great Lakes Water Quality Agreement (1972)

Lake Chad Basin Commission Agreement (1964)

Nile Basin Initiative (1999)

Transboundary Water Management in the Amazon Basin Agreement (2007)

Protocol on Water and Health (1999)

Columbia Basin Accord (2008)

Himalayan Rivers Treaty (proposed)

Sava River Basin Management Plan (2009)

Rio Declaration on Environment and Development (1992)

Helsinki Rules on the Uses of the Waters of International Rivers (1966)

UN Convention on the Law of Non-navigational Uses of International Watercourses (1997)

Danube River Protection Convention (1994)

Indus Waters Treaty (1960)

Southern African Development Community Protocol on Shared Watercourses (2000)

Framework Agreement on the Sava River Basin (2002)

Agreement on the Cooperation for the Sustainable Development of the Mekong River Basin (1995)

German Water Initiative for Central Asia (2008)

European Water Framework Directive (2000)

UN Watercourses Convention (1997)

Water Convention (1992)

Helsinki Rules (1966)

Danube River Protection Convention (1994)

Mekong Agreement (1995)

Ganges Water Sharing Treaty (1996)

African Water Vision 2025

Text B
Water Allocation and Sharing in the Chu-Talas River Basin

Pre-reading Questions

1. China is a collectivist culture which values sharing. Is sharing equally cherished in the Chu-Talas River riparian states like Kazakhstan and Kyrgyzstan?

2. In what ways do you think geopolitical factors might influence the management of a shared river basin?

3. What roles do you think international agreements could play in resolving water disputes between riparian states?

4. What diplomatic challenges do riparian states face in balancing economic development with environmental sustainability when managing shared rivers?

5. Can you provide examples of successful transboundary water diplomatic efforts? And do you know anything about diplomatic exchanges over the issue of Chu-Talas River Basin?

6. Why do issues like water pollution, over-harvesting, and climate change become a great concern for water diplomats working on issues of a transboundary river basin like the Chu-Talas River Basin?

7. What kinds of technical and infrastructural solutions can be utilized by water professionals in the diplomacy concerning shared waters like the Chu-Talas River Basin more effectively?

The Chu-Talas River Basin, shared by the Republic of Kazakhstan and the Kyrgyz Republic, is an example of cooperative management of transboundary waters in the conflictual setting of Central Asia. The Basin's administrative division and geographical arrangement have been key to fostering transboundary water cooperation.

The Basin, with the Chu and Talas Rivers located within the northern part of the Tian Shan and the eastern margin of the Turan lowland, lies in a semi-arid to arid continental climate zone characterized by cold winters and hot, dry summers.

The flow of the two rivers is highly seasonal with spring and summer snowmelt and glacier melt as the primary sources of run-off rather than precipitation. A large fraction of the approximately 2 million people living in the two catchments relies directly on the agrarian economy which is dependent on irrigation. The timely and sufficient availability of water for irrigation is critical for both the livelihoods of farmers and the stability of local societies in these areas.

In addition, the asymmetry in water accessibility and management responsibilities has driven both countries to seek the establishment of a legal framework for the joint operation of these infrastructure facilities. The mountainous upper basin of both rivers lies entirely within Kyrgyzstan. And all major water regulation facilities for the Chu and Talas Rivers, including dams, reservoirs, and canals, are in Kyrgyzstan, which controls the upstream of the rivers. However, Kazakhstan is located in the downstream plains. In other words, Kazakhstan depends on the operation and proper maintenance of those Kyrgyzstan facilities to ensure its water access during arid seasons to cater for its large irrigation needs.

In 2003, the governments of Kazakhstan and Kyrgyzstan took the initiative to start an international project "Promoting the Creation of the Commission for the Chu and Talas between Kazakhstan and Kyrgyzstan" with the engagement of water officials from the Ministry of Agriculture of the Republic of Kazakhstan and the Ministry of Agriculture, Water Resources and Processing Industry of Kyrgyzstan. The effort was endorsed by organizations including the Organization for Security and Co-operation in Europe (OSCE), the UN Economic Commission for Europe (UNECE), and the UN Economic and Social Commission for Asia and the Pacific (UNESCAP), with the participation of the Russian-Estonian Center of Transboundary Cooperation for the Chudskoye Lake. The project was launched in alignment with the UN Special Programme for the Economies of Central Asia (SPECA). The objective of the project was to provide assistance to Kazakhstan and Kyrgyzstan in implementing the Agreement in 2000 and to provide information to stakeholders and involve broad public participation into the management of water resources in the Chu-Talas River Basin.

Rounds of water diplomatic endeavors under the framework of the Project led to a bilateral Chu-Talas Rivers Commission in 2006. The Commission enables joint and transparent decision-making on water allocation and maintenance costs shared the two parties, as well as relevant data sharing, efficient implementation of joint projects, prevention and rapid settlement of problematic situations in the Chu-Talas River Basin.

In 2015, the Transboundary Water Commission for the Chu-Talas Rivers called for the release of quarterly hydrological reports and interactive online tools. One of its acting organs, the International Office for Water (IOW), offered technical and logistical support to gather the necessary data and produce those publications. Under the Commission's oversight, the bulletins are now regularly issued, and the interactive tools enable online access to data on river flow and the status of key water intakes for irrigation. The data in the reports are trustworthy as it is supplied by national experts and five institutions from both countries.

Access to More Working Knowledge for Better Communication

1. Read the facts from the UNESCO website below and engage in pairs or a mock negotiation session to discuss the importance of transboundary water management, evaluate the current status and progress of international efforts, and propose measures for its improvement.

—The report "Progress on Transboundary Water Cooperation. Global baseline for SDG indicator 6.5.2" shows that in some regions and basins, significant progress has been made to further transboundary water cooperation through operational arrangements. However, results from the 6.5.2 indicator monitoring exercise on transboundary water cooperation show that arrangements for transboundary water cooperation are often absent.

—The first SDG indicator 6.5.2 monitoring exercise has demonstrated that for the 62 countries considered, the average proportion of the transboundary basin area covered by an operational arrangement is 59 per cent, while only 17 countries have all their transboundary basins covered by operational arrangements. Cooperation on transboundary aquifers represents a particular challenge and is lagging further behind.

—Despite numerous services provided by groundwater for both humans and ecosystems, operational arrangements for transboundary aquifers are still rare around the world. Key issues that pose obstacles in realizing a commitment to coordinated or joint management of the "invisible resource" include a lack of adequate groundwater monitoring at national levels and detailed transboundary aquifer assessments.

—Basin-specific arrangements are greatly supported by regional and global frameworks such as the Convention on the Protection and Use of Transboundary Watercourses and International Lakes (Water Convention), the Convention on the Law of the Non-navigational Uses of International Watercourses (Watercourses Convention, which entered into force in 2014), the EU Water Framework Directive, and the Revised SADC Protocol on Shared Watercourse Systems. Where operational arrangements are lacking these instruments offer the tools upon which new arrangements can be negotiated or existing arrangements can be strengthened.

—UNESCO's International Hydrological Program also helps member states through in-depth assessments of transboundary aquifers, to establish a shared science-based understanding of the resource, and capacity building to support the process of institutionalizing cooperation among riparian countries. This led most recently to the establishment of a joint governance mechanism for the Stampriet Transboundary Aquifer System (STAS) between Botswana, Namibia and South Africa in 2017.

2. Read the following excerpt on transboundary cooperation in the Sava and Drina River basins about benefits of developing agreements or arrangements for transboundary waters and elaborate persuasive arguments to convince actors of transboundary water riparians to participate in regional cooperation.

The collaborative management of the Sava River basin, shared by Bosnia and Herzegovina, Croatia, Montenegro, Serbia, and Slovenia, exemplifies a "best practice" in transboundary cooperation, resulting in an effective process of socio-economic recovery in the basin through post-conflict cooperation over water (The Economist Intelligence Unit, 2019). The value of this cooperation is still evident today, as countries are jointly tackling emerging issues (notably climate change adaptation, including drought management) and strengthening cross-sectoral cooperation for sustainable planning and policy development, including the Drina River sub-basin, where most of the basin's hydropower is concentrated.

The International Sava River Basin Commission (ISRBC) was established in 2002 with the mandate of implementing the Framework Agreement for the Sava River Basin (FASRB). Remarkably, this was the first regional agreement to be signed since the Dayton Peace Agreement ended the war in the Yugoslavia. The restoration of inland navigation allowed the return of regional trade, strengthening economic integration across the countries and beyond, notably with the European Union. Rebuilding of bridges and ports throughout the basin accompanied the removal of war debris and mines, leading to the restoration of local livelihoods, including agriculture and tourism.

In the context of increasing tensions between different major water users, such as agriculture and energy, a participatory assessment of the water-food-energy-ecosystems nexus under the Water Convention was carried out in the Sava (2014) and later in the Drina (2016—2022, through multiple projects) River basin areas. The aim of these assessments was to look for cross-sectoral solutions to increase resource use efficiency, capitalize on regional complementarities, and improve natural resource governance.

These efforts resulted, among others, in the quantification of the benefits of transboundary cooperation on hydropower and the elaboration of possible ways to operationalize flow regulation in the basin (also through the establishment of a dedicated expert group), as part of a "nexus roadmap" for coordinating actions across sectors and countries. The roadmap aims to coherently guide policy-makers through the implementation of their sectoral and cross-sectoral strategic plans at the basin level (including notably through the Green Agenda for the Western Balkans GAWB). Climate adaptation, sustainable renewable energy planning, and sediment management are among the cross-sectoral activities included in the roadmap and also guide the "Sava and Drina Rivers Corridors Integrated Development Program".

Further Recommended Reading

[1] UN. ECE. *Policy guidance note on the benefits of transboundary water cooperation: identification, assessment and communication*. New York; Geneva: UN, 2015. https://digitallibrary.un.org/record/810004?v=pdf.

[2] Global High-Level Panel on Water and Peace. "*A matter of survival*", *the report of the Global High-Level Panel on Water and Peace draws a lot of attention*. 2017. https://www.genevawaterhub.org/news/matter-survival-report-global-high-level-panel-water-and-peace-draws-lot-attention.

[3] High-level Panel on Water. *Making every drop count: an agenda for water cooperation*. 2018. https://thedocs.worldbank.org/en/doc/623971522268005372-0090022018/original/17825HLPWOutcome.pdf

[4] Australian Government. *Murray-darling basin authority*. 2025. https://www.mdba.gov.au/.

Unit 6　Flood Prevention and Control

Background Study for Successful Intercultural Communication

Read authentic Chinese and English topical texts in advance before making a glossary for their associated terminology and jargon as follows.

pollution and flooding challenge 污染和洪水挑战

comprehensive water management plan 全面水管理计划

stricter environmental regulations 更严格的环境法规

China's construction of dams 中国的水坝建设

Associated Programme on Flood Management (of WMO and GWP) 洪水管理相关计划(世界气象组织与全球水伙伴)

Central European Flood Risk Assessment and Management 中欧洪水风险评估与管理

European Consortium on Meteorology Limited Area Ensemble Prediction System 欧洲气象联盟有限区域集成预测系统

State Hydraulic Works, Turkey 土耳其国家水务局

European Commission 欧洲委员会

European Centre for Medium-Range Weather Forecasts 欧洲中期天气预报中心

International Environmental Association of River Keepers 国际河流守护者环境协会

Eastern Europe, Caucasus and Central Asia 东欧、高加索及中亚

European Flood Awareness System 欧洲洪水预警系统

Environment and Security Initiative 环境与安全倡议

European exchange circle on flood forecasting 欧洲洪水预报交流圈

European exchange circle on flood mapping 欧洲洪水制图交流圈

flood awareness and prevention policy 洪水意识与预防政策

geographical information system 地理信息系统

Global Runoff Data Centre 全球径流数据中心

global system for mobile communications 全球移动通信系统

Global Water Partnership 全球水伙伴

International Commission for the Protection of the Rhine 保护莱茵河国际委员会

integrated flood management 综合洪水管理

International Network of Basin Organizations 国际流域组织网络

Joint Research Centre (of the European Commission) 欧洲委员会联合研究中心

Monitoring and Information Centre 监测与信息中心

New Approaches to Adaptive Water Management under Uncertainty 不确定性下的适应性水管理新方法

Organization for Security and Co-operation in Europe 欧洲安全与合作组织

Regional Environmental Center for Central and Eastern Europe 中东欧地区环境中心

United Nations Development Programme 联合国开发计划署

United Nations Economic Commission for Europe 联合国欧洲经济委员会

Convention on the Protection and Use of Transboundary Watercourses and International Lakes 跨界水道和国际湖泊保护与利用公约

World Hydrological Cycle Observing System 世界水循环观测系统

World Meteorological Organization 世界气象组织

Text A
Floods and Prevention in the Ganges-Brahmaputra-Meghna Basin

Pre-reading Questions

1. What are the primary causes of flooding in river basins, particularly in the Ganges-Brahmaputra-Meghna Basin?

2. How do different communities within the Ganges-Brahmaputra-Meghna Basin perceive flooding? Are there cultural narratives or beliefs associated with flooding events?

3. What historical events have shaped flood management policies in the Ganges-Brahmaputra-Meghna Basin? How have past floods influenced current practices?

4. In what ways do you think climate change is affecting flood patterns and prevention efforts in the Ganges-Brahmaputra-Meghna Basin?

The Ganges-Brahmaputra-Meghna (GBM) Basin, one of the world's most extensive river systems, is highly susceptible to flooding due to its unique hydrological and geographical characteristics. Spanning parts of India, Bangladesh, Nepal, Bhutan, and China, the Basin faces recurrent flood events that impact millions of people, disrupt socioeconomic activities, and cause substantial environmental damage. Effective flood prevention in the GBM Basin requires a comprehensive understanding of the Basin's dynamics, collaborative efforts among riparian countries, and the implementation of integrated flood management strategies. We strive to examine the flood prevention measures in the GBM Basin, drawing on factual cases to highlight the strategies and challenges involved.

Flooding in the GBM Basin is primarily driven by the monsoon rains, which contribute to high river flows and overflow of banks, particularly in the lower reaches. The flat topography and high population density in the floodplains exacerbate the vulnerability to floods. Bangladesh, located at the confluence of the Ganges, Brahmaputra, and Meghna rivers, is particularly prone to severe flooding, with significant portions of the country inundated during the monsoon season.

A notable example of flood prevention efforts in the GBM Basin is the construction of embankments and levees along the rivers. In India and Bangladesh, extensive networks of embankments have been built to contain river flows and protect agricultural lands and settlements. However, these structural measures have had mixed results. While they provide immediate protection, they can also lead to unintended consequences such as riverbed aggradation, reduced sediment deposition in floodplains, and increased flood risk in downstream areas.

One significant case is the Kosi River embankment project in Bihar, India. The Kosi River, known as the "Sorrow of Bihar", has a history of changing its course and causing devastating floods. To mitigate this, embankments were constructed in the 1950s. While these embankments initially provided flood protection, the river breached its embankments in 2008, resulting in one of the worst floods in the region's history. This event highlighted the limitations of relying solely on structural measures and underscored the need for more comprehensive flood management approaches.

Nonstructural measures, such as flood forecasting and early warning systems, play a crucial role in flood prevention in the GBM Basin. The establishment of the Flood Forecasting and Warning Centre (FFWC) in Bangladesh is a significant development in this regard. The FFWC uses hydrological and meteorological data to predict flood events and disseminate warnings to vulnerable communities. These warnings enable timely evacuation and preparedness measures, reducing the loss of life and property. However, challenges such as data accuracy, communication infrastructure, and community awareness continue to impact the effectiveness of early warning systems.

Transboundary cooperation is essential for effective flood prevention in the GBM Basin, given the interconnected nature of the river systems. The sharing of hydrological data among the riparian countries is critical for accurate flood forecasting and coordinated response. For instance, India and Bangladesh have been sharing real-time data on river flows during the monsoon season to improve flood forecasting capabilities. However, political sensitivities and trust deficits often hinder the full potential of such cooperative efforts.

The Brahmaputra River, with its vast catchment area and high sediment load, poses significant flood risks, particularly in Assam, India, and downstream in Bangladesh. Integrated flood management approaches, combining structural and nonstructural measures, are necessary to address these risks. In Assam, the construction of raised platforms and flood shelters provides refuge during flood events, while community-based flood management initiatives enhance local resilience. These initiatives include capacity building, awareness programs, and the promotion of traditional knowledge and practices in flood management.

In Bangladesh, the Comprehensive Disaster Management Programme (CDMP) aims to integrate disaster risk reduction into national development planning. The CDMP focuses on enhancing institutional capacities, improving early warning systems, and promoting community-based disaster preparedness. By involving local communities in flood management, the CDMP fosters a more inclusive and sustainable approach to flood prevention.

Climate change poses additional challenges to flood prevention in the GBM Basin, with increasing rainfall variability, glacier melt in the Himalayas, and rising sea levels contributing to more frequent and severe flood events. Adapting to these changes requires a dynamic and flexible approach to flood management. Measures such as the restoration of natural floodplains, reforestation in upstream areas, and the construction of multipurpose reservoirs can enhance the Basin's resilience to climate-induced flooding.

The restoration of wetlands and floodplains in the GBM Basin offers multiple benefits, including flood attenuation, groundwater recharge, and biodiversity conservation. Wetlands act as natural buffers, absorbing excess water during floods and releasing it during dry periods. Efforts to restore and protect wetlands, such as the Haor region in northeastern Bangladesh, contribute to sustainable flood management while supporting livelihoods and ecosystems.

Multipurpose reservoirs, designed for flood control, irrigation, and hydropower, provide a strategic approach to managing water resources in the GBM Basin. The construction of the Tehri Dam in Uttarakhand, India, exemplifies this approach. The dam's primary functions include flood control during the monsoon season and water storage for dry periods. However, the construction of large dams must consider potential socio-environmental impacts, including displacement of communities and changes in downstream river flow patterns.

Flood prevention in the Ganges-Brahmaputra-Meghna Basin necessitates a multifaceted approach that combines structural and nonstructural measures, fosters transboundary cooperation, and integrates climate change adaptation strategies. The cases of the Kosi River embankment, the FFWC in Bangladesh, and integrated flood management initiatives in Assam illustrate the diverse strategies employed to mitigate flood risks. Effective flood prevention requires not only technical solutions but also the active participation of local communities, the incorporation of traditional knowledge, and the commitment of riparian countries to collaborative water management. By embracing these principles, the GBM Basin can enhance its resilience to flooding and ensure sustainable development for its millions of inhabitants.

Access to More Working Knowledge for Better Communication

1. Read the list of actions and discuss in pairs or at a mock negotiating session about how water diplomats can facilitate regional flood prevention and control.

—Familiarize yourself with the causes of flooding, including meteorological factors (heavy rainfall, snowmelt), geographical features (topography, river basins), and human influences (urbanization, deforestation). This knowledge will help you communicate effectively about risks and mitigation strategies.

—Identify and engage with key stakeholders, including government agencies, local communities, NGOs, and environmental organizations. Understanding their interests and concerns is vital for fostering collaborative communication.

—Craft clear and concise messages that convey essential information about flood risks, prevention measures, and emergency response plans. Use simple language and visuals to enhance understanding.

—Encourage community participation in flood prevention efforts. Organize workshops, information sessions, and drills to raise awareness and gather input from local residents. Engaging communities fosters a sense of ownership and responsibility.

—Utilize technology and social media platforms to disseminate information quickly and efficiently. Tools like mobile apps, websites, and social media can facilitate realtime communication during flood events.

—Develop a crisis communication plan that outlines procedures for disseminating information before, during, and after a flood. This should include designated spokespersons, communication channels, and emergency contact information.

—Recognize cultural differences that may influence communication styles and perceptions of risk. Tailor your approach to respect local customs and languages, ensuring inclusivity and understanding.

—Establish channels for feedback from communities and stakeholders.

Listening to their concerns and suggestions can improve future flood prevention strategies and enhance trust.

—Implement educational programs that inform communities about flood risks, prevention techniques, and preparedness measures. Building knowledge and resilience can empower individuals and communities to respond effectively to flooding.

—Partner with hydrologists, urban planners, and disaster management experts to gather insights and develop comprehensive flood prevention strategies. Collaborative efforts ensure that communication is informed by scientific and technical expertise.

—Continuously monitor and evaluate flood prevention efforts and communication strategies. Assessing their effectiveness can identify areas for improvement and enhance future flood management initiatives.

2. Read more books on flood control and prevention and talk about the following arguments, discussing whether you are for or against them and why.

—Investing in flood prevention saves significantly more money in disaster recovery costs than the expenses incurred in implementing preventive measures.

—Proactive flood management enhances public safety and community resilience by ensuring preparedness for potential flooding events.

—Sustainable flood prevention strategies, like wetland restoration, improve biodiversity and water quality while mitigating flood risks.

—Collaborative flood management efforts foster social cohesion and trust among community members, enhancing overall preparedness.

—With climate change increasing flood frequency and severity, adopting forward-thinking flood prevention measures is essential for future adaptability. Early warning systems and effective communication strategies can significantly reduce panic and confusion during flooding emergencies.

—Proper land-use planning and zoning can minimize flood risk by preventing development in high-risk areas.

—Investment in green infrastructure, such as rain gardens and permeable pavements, can effectively manage stormwater and reduce flooding.

—Community education and awareness programs empower residents to take proactive measures for flood preparedness.

—International cooperation on flood management can lead to shared resources and knowledge, improving overall effectiveness in transboundary regions.

—Restoring natural floodplains allows rivers to absorb excess water, reducing downstream flooding impacts.

—Government policies that prioritize flood prevention can enhance economic stability by protecting vital infrastructure and industries.

—Engaging local stakeholders in flood management planning ensures that solutions are tailored to specific community needs and vulnerabilities.

—Climate-resilient infrastructure investments can withstand extreme weather events, reducing damage and recovery time.

—Comprehensive flood management strategies can create jobs and stimulate local economies through construction and maintenance projects.

Development of Intercultural Communication Skills

1. Transboundary water basins are home to a large number of people from different ethnic groups, religions, and cultural backgrounds. How do socioeconomic conditions of communities in a basin influence their vulnerability to flooding and their ability to implement prevention measures?

2. The Ganges-Brahmaputra-Meghna Basin spans multiple countries such as India, Bangladesh, and Nepal. Each country has its own political, cultural, and administrative systems. So what roles do local and national governments play in flood prevention and management in the Ganges-Brahmaputra-Meghna Basin? What challenges do they face?

3. In transboundary water basins, communities have rich and diverse cultural heritages. Some communities have traditional knowledge and experience in dealing with floods, such as the use of local materials to build temporary flood shelters. How important is community involvement in developing and implementing flood prevention strategies? Can you think of examples where community-led initiatives have been successful?

4. Different countries and regions in a basin have different levels of

technological acceptance and cultural attitudes towards engineering projects. What types of technological innovations or engineering solutions can be employed to prevent or mitigate flooding in river basins? What are the potential benefits and drawbacks of these technologies?

5. Communication and cooperation in flood prevention require overcoming language barriers and cultural differences. Given that the Ganges-Brahmaputra-Meghna Basin spans multiple countries, what challenges might arise in coordinating flood prevention efforts across national borders?

6. How can education and awareness programs contribute to improving disaster preparedness among communities prone to flooding in the basin?

7. What environmental factors must be considered when developing flood prevention strategies in the Ganges-Brahmaputra-Meghna Basin?

8. Community resilience is closely related to cultural traditions and social structures. In some communities, strong kinship and community organizations play an important role in coping with disasters. What are some effective strategies for building resilience among communities affected by frequent flooding in the basin?

9. Different countries and regions in the basin have different value orientations and cultural concepts, which will affect the assessment of flood management policies. Some regions may focus more on economic losses, while others may pay more attention to the impact on cultural heritage and social stability. How can we assess the effectiveness of existing flood management policies in the Ganges-Brahmaputra-Meghna Basin? What criteria should be used?

10. International collaboration in the basin requires mutual understanding and respect among different cultures. Through cultural exchanges and cooperation, countries can better understand each other's needs and positions in flood prevention and promote the smooth progress of cooperation. What potential benefits could arise from international collaboration on flood prevention in the Ganges-Brahmaputra-Meghna Basin?

11. Some traditional cultural beliefs may affect people's behavior and attitudes in the face of floods and health problems. Some religious groups may have specific rituals and taboos during disasters, which need to be considered when formulating public health measures. How do floods impact public health in the Ganges-Brahmaputra-Meghna Basin, and what measures can be taken to mitigate these effects?

Text B
Floods and Prevention Measures in the Danube River Basin

Pre-reading Questions

1. How do different countries along the Danube River perceive the impact of flooding on their communities and economies?

2. In what ways might cultural values influence attitudes toward flood prevention and management strategies in the Danube River Basin?

3. What role does local knowledge and traditional practices play in shaping flood prevention measures among diverse communities in the region?

4. How can cross-cultural collaboration enhance the effectiveness of flood management efforts in the Danube River Basin?

5. What are some challenges that arise when different countries or regions implement flood prevention policies, given their unique cultural and political contexts?

6. How do language barriers affect communication and cooperation among stakeholders involved in flood management in the Danube Basin?

7. In what ways can understanding local customs and beliefs improve the acceptance of flood prevention measures in different communities?

8. How do historical flood events shape collective memory and influence current flood management practices in the Danube River Basin?

9. What cultural factors might affect the willingness of communities to evacuate or respond to flood warnings?

10. How can education and awareness campaigns be tailored to address the specific cultural contexts of communities affected by floods in the Danube River Basin?

Flooding is the most common natural disaster in Europe and, in terms of economic damage, the most expensive one.

The Danube River Basin is Europe's second largest river basin, with a total area of 801,463 km^2. More than 80 million people from 19 countries share the Danube catchment area, making it the world's most international river basin. All countries comprising over 2,000 km^2 of the Basin, and also the European Union, are contracting parties of the International Commission for the Protection of the Danube River (ICPDR), which is charged with coordinating the conservation, improvement, and rational use of Danube waters.

Based on its gradients, the Danube River Basin can be divided into three sub-regions: the Upper, Middle, and Lower Basins (the latter including the Danube Delta). The Upper Basin extends from the source of the Danube in Germany to Bratislava in Slovakia. The Middle Basin is the largest of the three sub-regions, extending from Bratislava to the dams of the Iron Gate Gorge on the border between Serbia and Romania. The lowlands, plateaus, and mountains of Romania and Bulgaria form the Lower Basin of the River Danube. Before reaching the Black Sea, the river divides into three main branches, forming the Danube Delta, which covers an area of about 6,750 km^2.

The Tisza is the longest tributary of the Danube and the largest in terms of catchment area. At 966 km in length, it is the second largest in terms of flow after the Sava River. It drains an area of 157,186 km^2 in five countries (Slovakia, Ukraine, Hungary, Romania, and Serbia). The second longest river is the Prut (950 km), which is also the last major tributary of the Danube, with its confluence located just upstream of the Danube Delta. Further downstream, the Prut forms the border between Romania and Moldova. The second largest river in terms of catchment area, the Sava is 861 km in length and has the largest discharge of water to the Danube of any tributary. The Inn is the third largest of the Danube's tributaries in terms of discharge, and the seventh longest. Finally, the fourth largest and fourth longest Danube tributary is the Drava. Like many of the Danube rivers, the Drava has been considerably regulated with dams constructed to generate hydroelectricity and channels dredged to direct its flow. Nevertheless, natural habitats along the middle and lower reaches host unique assemblages of flora and fauna, as well as several endemic species.

Due to its large area and very diverse habitats—gravel islands on the Upper River, significant areas of forest floodplain, extensive wetlands on the Lower River—the Danube River provides the right living conditions in areas of high landscape and biodiversity for a large number of different species. Along its course, there are some 230 of the Danube River Basin's 2,860 Natura 2000 sites, an ecological network of internationally important protected areas in the territory of the European Union. Over 2,000 plant species and 5,000 animal species live in or by the waters of the Danube, a habitat which hosts about 2,000 vascular plants, over 40 mammals, approximately 100 fish species, as well as important bird sanctuaries for species such as the Dalmatian pelican. In addition, among the eight member states participating in the policy to protect these sanctuaries, Slovenia and Bulgaria have the highest terrestrial coverage rate of Natura 2000 sites in the whole EU. At 6,750 km^2, the delta of the Danube River is one of the world's largest wetlands (and Europe's largest remaining natural wetland) featuring rare fauna and flora, as well as 30 different types of ecosystems. Located in the territories of Romania and Ukraine, it became a UNESCO World Heritage Site in 1991.

The Danube countries all depend on the Danube waters as an economic resource. Despite this uniting feature, the Danube River flow through countries with very different social and economic conditions. For over forty years, Europe was divided into two political systems by the Iron Curtain. Following the collapse of the Soviet Union, the countries of Eastern Europe faced political and economic upheaval, with emerging free market democracies and the associated impacts of globalization, privatization, and deregulation. Simultaneously, industrial and agricultural output collapsed, leading to high unemployment. More recently, several Eastern European countries have made huge progress, successfully adapting production to international standards and acceding to the European Union. The socioeconomic data shows how diverse the region is. There is a wide gulf between the GDP per capita of Austria, Germany and Slovenia and the other Danube Basin countries: the wealthiest country's GDP per capita is nearly 14 times higher than that of the poorest. In terms of the characteristics of water use, data shows great differences in the Danube countries. These differences mainly correspond to the relative importance of the agricultural sector. While 9.8% of Ukrainian, 12.4% of Romanian, and 21.8% of Moldovan GDP are generated from agriculture, this share is only 1.7% for Austria, 2.4% for Germany, and 2.6% for the Czech Republic.

The economic value of the Danube River is immense; it is a vital resource for water supply, sustaining biodiversity, agriculture, industry, fishing, recreation, tourism, power generation, and navigation. A large number of dams, reservoirs, dykes, navigation locks, and other hydraulic structures have been built in the Basin to facilitate many of these important water uses. Historically, the Danube and some of its main tributaries, such as the Sava, have formed important trade routes across Europe for centuries. Moreover, the channelization of the river's course has made it easier for ships to navigate 2,411 km, or 87%, of the length of the Danube. As "Corridor VII" of the European Union, the Danube connects the Black Sea with the industrial centers of Western Europe and with the Port of Rotterdam. In recent years, there has been an increasing awareness for the need to balance economic and environmental needs in navigation management with special attention to the natural characteristics of the river. This was reflected in the "Joint Statement on Inland Navigation", issued by the ICPDR, Danube Commission, and Sava Commission. A similar consultation process is maintained between the ICPDR and representatives from the hydropower sector.

There have been 78 significant floods along the Danube over the last nine centuries; 23 of them took place in the 18th century before extensive flood protection works were started. Since then, significant areas of natural floodplains have been lost through drainage for agriculture, city development, and flood protection dykes—80% in total.

The curve of flood frequency has steepened in recent years, and high-water marks have set records three times since 2002. Five of the most significant floods have occurred in the last 10 years. Neglected levees have contributed to this damage, along with long winters and unusually heavy snow and rain. Multi-annual averages for precipitation have been exceeded by 1.5 to 2.0 times recently, reaching a maximum never before observed since systematic instrumental weather observations have been available.

The increasing regularity of dangerous hydrometeorological phenomena is a cause for concern. Estimation scenarios by the European Environmental Agency predict that flood damage and the number of people affected by flooding will rise substantially by 2100 as a result of climate change, with one scenario estimating an increase in flood damage of some 40% and an increase in the number of people affected by around 242,000 (about 11%). The EU formalized flood management in 2007 through the Floods Directive.

The Danube countries came together to sign the Danube River Protection Convention (DRPC) in 1994 and established the International Commission for the Protection of the Danube River (ICPDR) in 1998 to fulfill the Convention's objectives. The ICPDR is made up of 15 contracting parties (Austria, Bosnia and Herzegovina, Bulgaria, Croatia, the Czech Republic, Germany, Hungary, Moldova, Montenegro, Romania, Serbia, Slovakia, Slovenia, Ukraine, and the European Union) committed to implementing the DRPC; it is a forum for coordination and cooperation on important water management issues. The ICPDR comprises primarily national delegations that meet twice a year. With a secretariat based in Vienna, it is chaired by a president who serves for one year, and the presidency is passed on from one member country to another in alphabetical order. Much of the work of the ICPDR is done by Expert Groups, panels of specialists from the ICPDR member countries and 21 official observers. The ICPDR is also the facilitating platform for implementation of the EU Water Framework Directive (WFD) and the EU Floods Directive (EFD). Although not all members are part of the EU, contracting parties to the ICPDR have committed to meeting the goals of the WFD, which requires "good status" or "good potential" of all surface and ground waters by 2015.

Man-made changes to the natural course of the Danube waterways have interrupted river and habitat continuity, disconnected wetlands, and changed water quantity and flow conditions. Draining wetlands for agriculture often provides only marginal farmland while destroying unique wetland habitats. The introduction of foreign varieties of trees to floodplain forests and clear cutting in the name of industry eliminate undergrowth and alters the function of the floodplain ecosystem. Building towns and villages in floodplain areas also leaves them prone to damage from flooding. Species in the Danube River Basin are also threatened: migratory fish such as the Beluga have lost access to their spawning grounds due to dams and levees, threatening their existence in the Danube region.

In response to the danger of flooding, the ICPDR adopted the Action Programme on Sustainable Flood Protection in 2004. The goal of this program is to achieve a long-term and sustainable approach for managing the risks of flooding to protect human life and property, while encouraging conservation and improvement of water-related ecosystems.

Access to More Working Knowledge for Better Communication

1. Read the data listed below and discuss in pairs or at a mock negotiating session about the rationale for water diplomats to promote cooperative methods to disaster prevention and control in the world.

—1838 March Icy flood in Budapest. Damage: 10,100 houses, 153 lives lost.

—1888 spring Extreme floods in Tisza valley. Damage: Successful protection from dykes.

—1890 Major floods in the Upper Danube.

—1919 spring Extreme floods in Tisza valley. Damage: Successful protection from dykes.

—1925 Christmas Flood in the Krös valley, two dykes burst. 21,000 ha flooded, 904 houses destroyed.

—1954 Major floods in the Upper Danube.

—1965 April-July The biggest ever summer flood of the Danube at that time. Due to significant protection efforts a disaster was prevented. 11 dykes burst in the Rába valley.

—1970 May-July The largest ever Tisza valley flooding at that time. In other areas, the catastrophe was prevented by protection efforts. 14 dykes burst in the Romanian section of the Szamos and 3 in the Hungary; 57,000 ha flooded, 5,400 buildings destroyed; 9,000 inhabitants relocated for safety reasons.

—1992 Tazlău Romania. 107 lives lost.

—1997 Flooding on the Oder (Poland, Germany) and Morava/Danube (the Czech Republic, Germany). 105 lives lost.

—1998—2001 Extreme floods in four subsequent years at the Tisza.

—2002 Record flood levels for the Danube and the Elbe.

—2005 Record flood levels for the Danube. Romania: 31 lives lost, 55 roads closed, 600 bridges washed away. Bulgaria & Moldova: 14,000 people evacuated.

—2006 Record flood levels for the Danube. Bulgaria: 482 buildings destroyed, 300 people evacuated, 6,000 lives endangered. Romania: 300

buildings destroyed, 16,000 people evacuated, 150 villages flooded. Serbia: several thousand people evacuated due to a lack of clean water.

—2009 Record rain and flood levels for the Danube. Hungary: 198,000 ha affected. Romania: 110,000 ha of agricultural land were inundated by water.

—2010 Floods at Tisza and Danube simultaneously.

2. Use the following key words to expand your research on what aspects should be involved in water diplomacy to handle water-induced disasters and how to conduct intercultural communication with stakeholders on flood prevention and control.

flood risk assessment 洪水风险评估
floodplain management 洪泛区管理
early warning systems 预警系统
emergency response plan 应急响应计划
sustainable drainage systems 可持续排水系统
riverbank stabilization 河岸稳定
deforestation and flooding 砍伐森林与洪水
climate change impacts 气候变化影响
stormwater management 雨水管理
flood mitigation strategies 洪水缓解策略
levee construction 堤坝建设
floodplain restoration 洪泛区恢复
flood hazard mapping 洪水危险地图
public awareness campaigns 公众意识宣传活动
cross-border collaboration 跨境合作
disaster preparedness 灾害准备
infrastructure investment 基础设施投资
ecosystem-based approaches 基于生态系统的方法
floodplain zoning 洪泛区分区
flood insurance policies 洪水保险政策
riparian buffer zones 河岸缓冲区
sediment management 沉积物管理
flood recovery efforts 洪水恢复工作
watershed management 流域管理

rainwater harvesting 雨水收集
climate adaptation strategies 气候适应策略
flood early warning systems 洪水预警系统
ecological restoration 生态恢复
urban flood resilience 城市洪水韧性
flood control infrastructure 洪水控制基础设施
public-private partnerships 公私合营
flood emergency drills 洪水应急演练
risk communication 风险沟通
environmental impact assessment 环境影响评估
groundwater recharge 地下水补给
adaptive management practices 自适应管理实践

3. Read the excerpt of a UN paper below, thinking about how to implement water diplomacy to complement governmental diplomacy in the prevention of transboundary flood.

Transboundary river basins in the Hindu Kush Himalaya (HKH) are home to millions of people who depend on river water for their survival. Ten large Asian river systems that originate in the HKH—Amu Darya, Indus, Ganges, Brahmaputra, Irrawaddy, Salween, Mekong, Yangtse, Yellow, and Tarim—provide water and ecosystem services, and the basis for livelihoods to around 240 million people living in the mountains. Together, these river basins provide water to 1.9 billion people, a fifth of the world's population. The people of these river basins live in a multi-hazard environment, where floods, landslides, and droughts affect them every year. Disasters in the region often have cascading transboundary impacts with upstream downstream linkages. While floods may originate in the upstream, their impacts are also experienced downstream. In addition, an increase in the incidence of extreme weather events due to climate change and environmental degradation is expected to magnify the frequency and impact of such disasters. Women, children, and poor and marginalized communities are most vulnerable as they lack access to information and capacity to prepare for disasters and deal with their aftermath.

The Koshi and Gandaki river basins contain rich biodiversity and are a source of valuable ecosystem services that sustain the lives and livelihoods of millions of people in China, India, and Nepal. The basins play a key role in the irrigation of downstream areas and have great potential for hydropower development. However, the basins are highly prone to erosion, sedimentation, and natural hazards, which may increase in magnitude and frequency in the current context of climate change. Increasing population, urbanization, and encroachment put additional pressures on the basin's freshwater ecosystems. There are numerous grassroots organizations and practitioners (both in upstream and downstream reaches of Nepal and India) working to address disaster risk reduction in the Koshi and Gandaki river basins. There is a shared understanding of common issues, and this provides an opportunity for transboundary cooperation to address key threats such as floods, which cause loss of lives and livelihoods in the region. Presented here are major points of discussion that emerged from a transboundary dialogue between people working in Nepal and India on the issue of river basin management. Causes and common myths around floods in the basins are: i) embankments provide assured long-term protection from floods, ii) Nepal releases water that causes flood in Bihar. Those myths around floods are sociopolitical constructions on both sides of the border (Nepal and India), perpetuated by power play, a fatalist mentality, and the blame game.

Further Recommended Reading

[1] Biswas B, Ghute B B. *Flood risk management: assessment and strategy*. Singapore: Springer, 2024.

[2] Schmitt, A., and Werners, S. E. *Flood risk management in Europe: innovation in policy and practice*. Netherlands: Wiley, 2016.

[3] Edmund C. Penning Rowsell, Matilda Becker. *Flood risk management: global case studies of governance, policy and communities*. London: Routledge, 2019.

[4] United Nations Office for Disaster Risk Reduction. *2015 global assessment report on disaster risk reduction*. Geneva: UNISDR, 2015. https://digitallibrary.un.org/record/4037965? v=pdf.

[5] World Meteorological Organization. *Guidelines on flood management*. Geneva: WMO, 2013.

Unit 7 Drought Relief

Background Study for Successful Intercultural Communication

Read authentic Chinese and English topical texts in advance before making a glossary for their associated terminology and jargon as follows.

riparian states 流域国家
water allocation 水资源分配
water scarcity 水资源短缺
geopolitical landscape 地缘政治格局
semi-arid climate 半干旱气候
water demand 水需求
National Water Carrier Project (Aqaba-Amman Water Desalination and Conveyance Project) 国家输水工程项目(亚喀巴—安曼海水淡化和输送项目)
Sea of Galilee 加利利海
coastal plain 沿海平原
Negev Desert 内盖夫沙漠
East Ghor Canal 东戈尔运河
King Abdullah Canal 阿卜杜拉国王运河
Yarmouk River 亚尔穆克河
unilateral actions 单边行动
Joint Water Committee 水资源联合委员会
dialogue and coordination 对话与协调
drought contingency plans 干旱应急计划
Red Sea-Dead Sea Water Conveyance Project 红海死海输水项目
declining water levels 水位下降
desalination plants 海水淡化厂
shared water challenges 共享水资源挑战

disrupted water infrastructure 水基础设施被破坏
displaced millions 数百万人流离失所
strained water resources 水资源紧张
influx of Syrian refugees 叙利亚难民涌入
international assistance 国际援助
improved water use efficiency 提高用水效率
investing in water infrastructure 投资水基础设施
United Nations 联合国
World Bank 世界银行
Palestinian territories 巴勒斯坦领土
West Bank 西岸
Israeli-Palestinian conflict 以巴冲突
restricted access 受限的获取
political and security concerns 政治和安全问题
trilateral water agreement 三边水协议
increased water allocations 增加的水分配量
water infrastructure projects 水基础设施项目
collaborative drought relief efforts 合作性干旱救济努力
equitable water-sharing arrangements 公平的水分配安排
cooperative water management 合作水资源管理
resilient water management practices 有韧性的水资源管理实践

Text A

The Colorado River Basin Running Dry

Pre-reading Questions

1. Some people say precipitation is a key factor of drought. Are you on the same page with them? Do people from American culture agree with them? And why or why not?

2. Do you think high temperatures, increased evaporation, and human activities like water overuse and poor land management can also induce droughts? Have you ever shared your ideas with people from other cultures?

3. Do you think droughts only affect rural areas, or can they also significantly impact urban communities and infrastructure?

4. How do you understand droughts as events which might sometimes last for years or even decades affecting ecosystems and economies in the long run?

5. Could your region be at risk for droughts, even though it has historically had abundant rainfall? How might climate change bring about this risk?

6. Is drought simply a natural cycle, or are climate change and human activities making their frequency and severity more unpredictable?

7. Can water diplomats rely solely on technology to manage droughts, or might it require substantial investment and still not fully address the challenges posed by severe droughts?

Owing to climate change, severe droughts are expected to increase in both intensity and frequency. The Colorado River Basin, shared between the United States and Mexico, is a crucial and highly regulated water system in North America. It originates in the Rocky Mountains and flows through seven western U.S. states—Wyoming, Utah, Colorado, Arizona, California, Nevada, and New Mexico—before reaching the Gulf of California in Mexico, forming the border between the Mexican states of Sonora and Baja, California. Over 98% of the basin's area is within the U.S., and nearly all runoff is generated there, primarily sustained by snowmelt, with the majority of its flow sourced upstream of the Grand Canyon.

The basin is intricately managed for hydropower and water distribution, with numerous large dams enabling artificial storage capacities that exceed the basin's annual flow by more than four times. Approximately 70% to 80% of the water in the U.S. portion of the basin is allocated to agriculture, while the remainder supports municipal, industrial, recreational, and other uses.

A persistent imbalance between water supply and demand makes the system highly vulnerable to climate change. Rising temperatures and decreasing runoffs are expected to exacerbate drought severity, which is already an escalating challenge in the lower basin states.

Several national and international agreements govern the allocation and management of the Colorado River Basin, with the most significant being the

1922 Colorado River Compact and the 1944 U. S. -Mexico Treaty. The national agreements primarily focus on water distribution, with the 1922 Compact addressing allocations among seven U. S. states. The Compact allows for some flexibility in water allocations between the Upper and Lower Basins. However, it does not directly address drought conditions.

Internationally, the 1944 Treaty includes provisions for cooperation between the U. S. and Mexico, notably through the establishment of the International Boundary and Water Commission. Article 10 of the 1944 Treaty acknowledges the possibility of extraordinary droughts, permitting proportional reductions in water deliveries. Despite this, the Treaty lacks clarity on the criteria for determining allocations during droughts. Over time, amendments through legally binding "minutes" have been added to address evolving issues. For instance, Minute 319 in 2012 introduced water conservation, infrastructure development, and environmental measures for binational water management, and Minute 323 in 2017 created a binational water scarcity contingency plan.

Most of the water conservation efforts to mitigate supply challenges have been focused on the Lower Basin, where usage surpasses available allocation. Water consumption remains well below allocated levels in the Upper Basin. If a prolonged basin-wide drought occurs, adjustments to consumptive use from the Upper Basin to the Lower Basin may be necessary, although the current flexibility mechanisms established through minutes offer room for adaptation in response to increased variability.

The 1944 Treaty has also set a precedent for the sharing of flow data. In addition, other types of information exchange, such as water quality monitoring, have occurred through subsequent agreements. However, the focus on data exchange has largely been on ensuring water quality rather than addressing climate change's impacts on water resources. Despite this, monitoring data have proven valuable in shaping current adaptation plans.

Regarding funding for adaptation activities, no comprehensive basin-wide funding initiatives exist. However, temporary programs like the Pilot System Conservation Program, launched in 2014, test conservation measures to maintain reservoir levels. These measures have produced limited water savings in comparison to the total annual flow. Moving forward, further investments in conservation are expected to be needed to enhance the basin's resilience to climate variability.

Both Mexico and the U. S. are aware of the growing challenge of climate change. Mexico's National Water Commission has developed a National Drought Program, focusing on forecasting, coordination, and multi-level prevention efforts. Conversely, the U. S. lacks a national climate change policy but has implemented adaptive mechanisms, especially in the Lower Basin, where urban water shortages are more acute. Arizona, for example, has invested in groundwater banks to mitigate short-term water shortages. Additionally, the U. S. has introduced policies to encourage conservation, such as the 2007 Interim Guidelines for managing Lake Mead and Lake Powell during drought and low-water conditions.

Although significant steps have been taken, the lack of basin-wide vulnerability assessments limits the ability to plan for long-term climate impacts. Collaborative efforts, including the Bureau of Reclamation's 2012 water demand study and other planning initiatives, are necessary to determine viable solutions for meeting future water supply demands as the effects of climate change intensify.

Access to More Working Knowledge for Better Communication

Read the facts on drought and discuss in pairs or at a mock negotiating session about how drought can affect everyone's life and convince people to actively work as water diplomats within their own field to promote drought prevention and relief.

—Drought is typically categorized into four types, each with distinct causes and impacts. Meteorological drought occurs when there is insufficient rainfall or precipitation, while agricultural drought is triggered by a lack of soil moisture, disrupting crop growth. Hydrological drought arises from diminished water levels in lakes, rivers, and other water sources. Lastly, socioeconomic drought affects communities through shortages in drinking water and tap water supplies, impacting human well-being directly.

—An unsettling fact about droughts is that they are much harder to predict than other sudden weather phenomena, such as hurricanes, thunderstorms, or tornadoes. The effects of a drought often take weeks or even months to manifest, and determining when a drought will end is even more uncertain.

—Climate change has already influenced global weather patterns, intensifying droughts and increasing the variability of hydroclimate trends. These abnormal patterns are expected to persist and worsen as global temperatures continue to rise.

—NASA's groundbreaking research leverages tree ring analysis to understand historical droughts and climate patterns. By combining this data with modern soil moisture observations, scientists are better equipped to estimate future drought risks in the 21st century. One key finding from NASA's research reveals that human activity has significantly influenced global drought patterns for nearly a century.

—Droughts can also result in devastating consequences for agriculture. During severe dust bowl events, topsoil is completely stripped away by the wind, leaving behind nutrient-poor soil that cannot sustain crops, even long after the drought has ended.

—In more recent history, Texas experienced its longest drought since 2000, which lasted from May 2010 to July 2015. The most severe period occurred during the week of October 4, 2011, highlighting the intensity of this event.

—Drought doesn't just affect the environment—it can also exacerbate social tensions and conflicts. One stark example is the Darfur crisis in Sudan. Decades of drought, combined with overpopulation and land degradation, forced people to search for water resources beyond their regions. This led to territorial disputes and, ultimately, armed conflict.

Development of Intercultural Communication Skills

1. In water diplomacy, it is essential to respect different cultures. As future water professionals we have to study riparian states' cultural traditions and societal norms in tackling droughts. Are you for or against the statement? And why?

2. Different cultures may have varying beliefs about the causes of drought. In

some indigenous cultures, drought may be seen as a spiritual punishment. How can relief workers from Western science-based cultures communicate with these communities to provide practical drought-relief measures while respecting their cultural beliefs?

3. In certain cultures, there is a strong value placed on self-reliance and family-centered support systems. How can international drought-relief organizations communicate the benefits of external aid and cooperation to these communities without undermining their cultural values?

4. In a drought-affected region with multiple languages spoken, what are the most effective strategies for relief workers to communicate essential information about water conservation, rationing, and access to emergency water sources?

5. Some languages may lack specific terms related to modern drought-relief technologies or concepts such as desalination or cloud-seeding. How can relief workers accurately convey these ideas to non-native speakers?

6. In some cultures, there are strict gender roles where women may be responsible for water collection. How can drought-relief programs communicate and implement changes in water management strategies while respecting these gender norms?

7. In hierarchical societies, decision-making regarding drought-relief may be concentrated at the top levels. How can grassroots organizations and international NGOs communicate their ideas and needs to the relevant authorities in a way that respects the local social structure?

8. Different cultures may have different levels of trust in various media sources. In some regions, traditional word-of-mouth may be more trusted than modern media like radio or the Internet. How can drought-relief information be effectively disseminated using multiple media channels to reach diverse cultural groups?

9. In cultures where visual communication is highly valued, how can drought-relief campaigns create effective visual materials such as infographics or short videos that accurately convey complex information while being sensitive to cultural taboos and aesthetics?

Text B
Droughts in the Jordan River Basin

Pre-reading Questions

1. How can understanding the historical context of water disputes in the Jordan River Basin help stakeholders appreciate the complexity of current drought challenges?

2. How can intercultural communication skills help stakeholders navigate the political and cultural complexities of conflicts, such as the Israeli-Palestinian conflict, that hinder water management and drought relief?

3. How can mechanisms like the Joint Water Committee be used as platforms not only for technical coordination but also for fostering deeper intercultural understanding and mutual respect among the participating states?

The Jordan River Basin, a vital water source for Israel, Jordan, Lebanon, Palestine, and Syria, epitomizes the intricate interplay of cooperation and conflict in transboundary water management. The Basin's limited water resources are strained by high population densities, agricultural demands, and industrial activities, aggravating the impact of recurring droughts. Effective drought relief in the Jordan River Basin requires a blend of collaborative efforts and conflict resolution mechanisms among the riparian states. This case tells the complexities of drought relief in the Basin, drawing on facts to illustrate the dynamics of cooperation and conflict.

Drought conditions in the Jordan River Basin have historically led to tensions over water allocation and usage. The region's semi-arid climate, coupled with increasing water demand, has made water scarcity a chronic issue. One of the most prominent conflicts arose from the diversion projects undertaken by Israel and Jordan in the mid-20th century. Israel's National Water Carrier Project, completed in 1964, aimed to transport water from the Sea of Galilee to the coastal plain and Negev desert. This diversion significantly reduced the flow of the Jordan River downstream, impacting Jordanian agriculture and water supply.

In response, Jordan initiated the East Ghor Canal (now known as the King Abdullah Canal) to utilize water from the Yarmouk River, a tributary of the Jordan River. These unilateral actions heightened tensions and underscored the need for cooperative water management. The situation was further complicated by the geopolitical landscape, including the Israeli-Palestinian conflict and broader Arab-Israeli tensions.

Despite these conflicts, there have been notable instances of cooperation in drought relief and water management. The 1994 Israel-Jordan Peace Treaty marked a significant milestone in transboundary water cooperation. The Treaty included provisions for the equitable distribution of water from the Jordan and Yarmouk rivers and established mechanisms for joint water management. One key element of the Treaty was the establishment of the Joint Water Committee, which facilitates dialogue and coordination on water issues, including drought relief.

The Joint Water Committee has enabled both countries to share data, develop drought contingency plans, and undertake joint projects to enhance water supply and efficiency. For instance, Israel has provided Jordan with additional water from the Sea of Galilee during severe drought periods. This cooperative approach has alleviated some of the tensions and contributed to more resilient water management practices in the Basin.

Another significant case of cooperation is the Red Sea-Dead Sea Water Conveyance Project, involving Israel, Jordan, and Palestine. This ambitious project aims to address the declining water levels of the Dead Sea while providing potable water and supporting drought relief efforts. The project involves constructing a pipeline to convey water from the Red Sea to the Dead Sea, with desalination plants along the way to produce fresh water for the region. The project represents a collaborative effort to tackle shared water challenges and mitigate the impacts of drought on the basin's riparian states.

In contrast to these cooperative efforts, the Syrian conflict has exacerbated water scarcity and hindered collaborative drought relief in the northern part of the Jordan River Basin. The conflict has disrupted water infrastructure, displaced millions, and strained the water resources of neighboring countries, particularly Jordan and Lebanon. The influx of Syrian refugees into Jordan has significantly increased water demand, placing additional pressure on already scarce resources and complicating drought relief efforts.

To address the heightened water stress, Jordan has implemented various measures, including improving water use efficiency, investing in water infrastructure, and seeking international assistance. International organizations such as the United Nations and the World Bank have played a crucial role in supporting Jordan's efforts to manage water scarcity and provide drought relief. However, the lack of a comprehensive regional framework for managing the Basin's water resources continues to pose a challenge.

The Palestinian territories, particularly the West Bank, also face significant water scarcity issues worsened by drought conditions. The Israeli-Palestinian conflict complicates water management and access, with Palestinians often experiencing restricted access to water resources. Efforts to enhance cooperation in water management are hindered by political and security concerns. Nonetheless, initiatives such as the trilateral water agreement between Israel, Jordan, and Palestine in 2013 aimed to improve water supply and address some of the region's pressing water challenges. This agreement included provisions for increased water allocations and cooperation on water infrastructure projects, highlighting the potential for collaborative drought relief even in a complex geopolitical context.

The Jordan River Basin exemplifies the dual nature of transboundary water management, where cooperation and conflict coexist in the pursuit of drought relief. Factual cases such as the Israel-Jordan Peace Treaty, the Red Sea-Dead Sea Water Conveyance Project, and various bilateral and trilateral agreements demonstrate the potential for collaborative approaches to managing water scarcity. However, ongoing conflicts, geopolitical tensions, and disparities in water access continue to pose significant challenges. Effective drought relief in the Jordan River Basin requires sustained dialogue, equitable water-sharing arrangements, and the active involvement of international stakeholders to support and facilitate cooperative water management. By learning from past experiences and fostering a spirit of collaboration, the riparian states may work towards a more resilient and sustainable future for the Basin's water resources.

Access to More Working Knowledge for Better Communication

1. Read the facts listed below and discuss in pairs or at a mock negotiating session about how we can mobilize action in order to promote water diplomacy to contribute to reasonable water allocation in transboundary water basins.

—Transboundary river basin drought emergencies caused by extreme climate and human activities can increase the risk perception of decision makers, causing them to overestimate the consequences of disasters, which could intensify the water competition among riparian countries.

—Authorities have not given sufficient priority to equity and non-discrimination in access to Water, Sanitation and Hygiene (WASH) services, particularly between formal and informal settlements, rural and urban areas, highest and lowest wealth quintiles, and among marginalized groups.

—Water insecurity can be a major driver of conflicts and displacement, particularly in regions where water resources are limited or unevenly distributed.

—Inequalities in access to water and sanitation services, while not necessarily a direct driver of conflict, present significant barriers to socioeconomic stability and prosperity. WASH, as a "politically neutral" service system, can be leveraged as a platform for intercommunal collaboration and partnerships between citizens and governments.

—Numerous challenges undermine the provision of WASH services in conflict situations. Conflict can disrupt water supply systems, leading to damaged or non-functioning water infrastructure. Contaminated water sources and the absence of proper sanitation facilities increase the risk of waterborne diseases.

—Conflict-related displacement frequently leads to overcrowded living conditions in camps or informal settlements, which often lack adequate sanitation facilities, posing a great risk to the health and well-being of the most vulnerable, especially women, children, and persons with disabilities.

—Lack of water security, for example, has been identified as one of the key drivers of migration, as it undermines people's everyday existence and livelihood opportunities.

2. Explore the drought-specific databases and portals featured in the Global Drought Snapshot 2023 to access valuable resources that can support the development of effective drought preparedness strategies. These tools are essential for making informed decisions and enhancing your capabilities as future water diplomats.

—United Nations Convention to Combat Desertification (UNCCD)- Drought Toolbox https://knowledge.unccd.int/drought-toolbox

—The International Disaster Database https://www.emdat.be

—United Nations Office for Disaster Risk Reduction (UNDRR) PreventionWeb-Drought solutions https://www.preventionweb.net/collections/drought-solutions

—EDO-European Drought Observatory/GDO-Global Drought Observatory https://drought.emergency.copernicus.eu/tumbo/edo/map/

—FAO Drought portal https://www.fao.org/land-water/water/drought/drought-portal/en/

Further Recommended Reading

[1] International Boundary and Water Commission. *Minute No. 323: extension of cooperative measures and adoption of a binational water scarcity contingency plan in the Colorado River Basin*. 2019. https://www.ibwc.gov/wp-content/uploads/2023/03/Min323.pdf.

[2] Stern, C. V., and Sheikh, P. A. *Management of the Colorado River: water allocations, drought, and the federal role*. Congressional Research Service Report, 2025. https://www.congress.gov/crs_external_products/R/PDF/R45546/R45546.41.pdf.

[3] Blumstein, S., and Petersen-Perlman, J. D. When the water runs dry: supporting adaptive governance in transboundary river basins. *Water International*, 2021, 46(3): 306–324.

[4] Tsegai, D., Augenstein, P., Huang, Z., et al. *Global drought snapshot 2023: the need for proactive action*. 2023. https://reliefweb.int/report/world/global-drought-snapshot-2023-need-proactive-action-enru.

Unit 8 Water Pollution Prevention and Control

Background Study for Successful Intercultural Communication

Read authentic Chinese and English topical texts in advance before making a glossary for their associated terminology and jargon as follows.

nutrient pollution 营养物污染
heavy metal contamination 重金属污染
aquatic ecosystems 水生生态系统
water scarcity 水资源短缺
toxic substances 有毒物质
environmental assessment 环境评估
cross-border pollution 跨境污染
groundwater contamination 地下水污染
conservation strategies 保护策略
water footprint 水足迹
ecological restoration 生态恢复
habitat destruction 栖息地破坏
legislative frameworks 立法框架
source water protection 水源水保护
clean water initiatives 清水倡议
water safety plans 水安全计划
risk assessment 风险评估
transboundary water conflicts 跨界水冲突
local community involvement 当地社区参与
pollution prevention strategies 污染预防策略
bioremediation 生物修复
water resource allocation 水资源分配

climate adaptation 气候适应
public awareness campaigns 公共意识宣传
source pollution reduction 源头污染减少
intergovernmental collaboration 政府间合作
environmental justice 环境正义
integrated monitoring 综合监测
transboundary water issues 跨界水问题
environmental degradation 环境退化
industrial discharge 工业排放
agricultural runoff 农业径流
wastewater treatment 废水处理
pollution control measures 污染控制措施
water quality monitoring 水质监测
ecological impact 生态影响
sustainable water management 可持续水管理
ecosystem health 生态系统健康
public health risks 公共健康风险
pollutant load 污染物负荷
restoration efforts 恢复努力
pollution abatement 污染消减
waterborne diseases 水源性疾病
sediment contamination 沉积物污染
bioaccumulation 生物累积
transboundary monitoring programs 跨界监测项目
pollution hotspots 污染热点
water quality standards 水质标准

Text A
Rehabilitation Efforts in the Jordan River Basin

Pre-reading Questions

1. What challenges might arise when negotiating pollution control measures for a transboundary river that serves communities with diverse cultural, political,

and economic contexts?

2. How do socio-political relationships between countries sharing the Jordan River affect their willingness and ability to cooperate on water pollution prevention and management?

3. How can grassroots initiatives, such as EcoPeace's Middle East's Good Water Neighbors project, empower communities from different cultural and political backgrounds to contribute to river restoration and sustainable management?

4. What lessons from the Jordan River's history and current management strategies can be applied to other transboundary river management, given the need for culturally sensitive and inclusive approaches?

5. What do you think necessitate continued cooperation and innovation in water diplomacy while significant strides have been made in improving water quality?

The Jordan River, a critical watercourse in the Middle East, holds immense cultural, religious, and geopolitical significance. It flows from the Anti-Lebanon Mountains through the Sea of Galilee and down to the Dead Sea, traversing Israel, Jordan, and the Palestinian territories. Historically, the river has been a lifeline for agriculture and human settlements in the region. However, over the past century, the Jordan River has experienced severe water pollution, prompting significant efforts to implement water pollution prevention and control measures.

The River's pollution history is closely tied to regional socio-political dynamics, rapid population growth, and agricultural intensification. Beginning in the early 20th century, increasing agricultural activities and urbanization led to higher water withdrawals and the discharge of untreated effluents into the river. The establishment of large-scale irrigation projects, such as the National Water Carrier of Israel and the East Ghor Canal in Jordan, significantly diverted water from the river, reducing its flow and concentrating pollutants.

By the mid-20th century, the Jordan River's water quality had markedly deteriorated. Untreated sewage, agricultural runoff containing pesticides and fertilizers, and industrial waste significantly polluted the river. The lower Jordan River, in particular, became increasingly saline and laden with pollutants, severely affecting its ecological health and usability for agricultural and domestic purposes.

Initial attempts to address pollution in the Jordan River were limited and largely ineffective due to the region's complex political landscape and the prioritization of water quantity over quality. Fragmented management and lack of coordination among riparian countries hindered comprehensive pollution control.

Significant efforts to improve water quality in the Jordan River began in the late 20th and early 21st centuries, driven by increasing awareness of the environmental and health impacts of pollution and a growing recognition of the need for regional cooperation.

One of the earliest comprehensive initiatives was the multilateral working group on water resources established during the Madrid Peace Conference in 1991. This group aimed to foster regional cooperation on water management including pollution control. However, political tensions often impeded the progress.

More concrete measures were implemented through various bilateral and multilateral agreements and projects. Key initiatives included:

The Red Sea-Dead Sea Project: This ambitious project, initiated by Jordan, Israel, and the Palestinian Authority, aims to address both water scarcity and pollution in the Jordan River Basin. By transferring water from the Red Sea to the Dead Sea, the project seeks to replenish the Jordan River and mitigate pollution through improved wastewater treatment and agricultural practices.

The Good Water Neighbors Project: Launched by EcoPeace Middle East, this project fosters cross-border cooperation on water management by bringing together communities from Israel, Jordan, and Palestine. The project promotes sustainable water use practices, pollution reduction, and joint advocacy for river restoration.

Wastewater Treatment Upgrades: Significant investments have been made in upgrading wastewater treatment facilities in Israel, Jordan, and the Palestinian territories. Improved treatment processes have reduced the discharge of untreated sewage into the Jordan River, thereby enhancing its water quality.

Agricultural Best Practices: Efforts to promote sustainable agricultural practices have been implemented to reduce the runoff of pesticides and fertilizers into the river. Those include training farmers in integrated pest management, efficient irrigation techniques, and the use of organic fertilizers.

The implementation of these measures has led to some improvements in the Jordan River's water quality. Upgraded wastewater treatment facilities have reduced the volume of untreated sewage entering the river, and sustainable agricultural practices have helped decrease the runoff of harmful chemicals. However, the river's flow remains severely diminished due to extensive water withdrawals, and the pollution levels are still high compared to historical baselines.

Therefore, the cooperative efforts among riparian countries and various stakeholders have laid a foundation for further progress. The integration of pollution control with broader water management strategies, such as the Red Sea-Dead Sea Project, exemplifies the holistic approach needed to address the complex issues facing the Jordan River.

The Jordan River's experience underscores several critical lessons for water pollution prevention and control. First, regional cooperation is essential for managing transboundary water resources effectively. The river's rehabilitation efforts have demonstrated that collaborative projects and joint management initiatives can yield positive outcomes, even in politically complex regions. Second, a holistic approach that integrates water quantity and quality management is vital. Addressing pollution in the Jordan River requires not only improving wastewater treatment and agricultural practices but also ensuring sufficient flow to dilute pollutants and maintain ecological health.

Looking forward, the Jordan River faces ongoing challenges. Climate change, population growth, and continued political tensions pose significant risks to water quality and availability. Strengthening regional cooperation, enhancing adaptive management strategies, and investing in innovative technologies for pollution control will be crucial to sustaining and building on the progress made.

Access to More Working Knowledge for Better Communication

Read the facts listed below and discuss in pairs or at a mock negotiating session about how we can make a difference by trying to act as a water diplomat in our own field to alleviate water pollution.

—Worldwide, agriculture accounts for roughly 70% of freshwater withdrawals, followed by industry (just under 20%) and domestic (or municipal) uses (about 12%).

—Groundwater supplies about 25% of all water used for irrigation and half of the freshwater withdrawn for domestic purposes.

—Since 1980s, the global demand for freshwater has been increasing by just under 1% per year.

—Water demand from the municipal sector has experienced a considerable increase relative to the other sectors, and is likely to continue growing for population urbanization and the expanding of the water supply and sanitation systems servicing these cities.

—In lower-income countries, poor water quality is due to low levels of wastewater treatment, whereas in higher-income countries, runoff from agriculture poses the most serious problem.

—Emerging contaminants include pharmaceuticals, hormones, industrial chemicals, detergents, cyanotoxins, per- and poly-fluoroalkyl substances (PFAS) and nanomaterials.

Development of Intercultural Communication Skills

1. While you can lead a horse to water, you can't make it drink—true collaboration requires willingness and commitment from all parties. At COP27 in 2022, Israel and Jordan demonstrated a great commitment by announcing their joint efforts toward the ecological rehabilitation, restoration, and sustainable development of the Jordan River. How might such a collaborative initiative be used by experienced water professionals?

2. The relationship between Israel and Jordan is marked by a history of tension, despite their peace treaty signed in 1994. The willingness to cooperate in restoring the Jordan River demonstrates a pragmatic approach to building trust through shared interests. What should water professionals in the cooperation pay special attention to in overcoming the political and historical challenges?

3. Cultural sensitivity involves the awareness, understanding, and respect for the differences and diversity in values, beliefs, traditions, and practices across cultures. In a meeting between stakeholders from two countries with distinct historical and cultural backgrounds, how can water diplomats mediate disputes and

foster cooperation in preventing water pollution to achieve a fair and mutually beneficial compromise?

4. The patterns of power distribution and interaction between groups, entities or countries shape how influence, authority, and decision-making responsibilities are allocated and exercised in a water pollution incident. What should future water diplomats be mindful of when studying how to work in an imbalanced power dynamic?

5. What steps should a qualified water diplomat take to address the disparities if he or she recognizes any in economic capacity among stakeholders?

Text B

The Rhine River: From Pollution Crisis to Environmental Rehabilitation

Pre-reading Questions

1. The Rhine River has seen the thriving development of different cultures on both sides of its banks. It has served as a channel for the spread of cultures, religions, and artistic styles, enabling the mutual influence and integration of diverse cultures. What more do you know about the Rhine River?

2. Cooperative models like International Commission for the Protection of the Rhine have promoted political trust and cooperation among countries and laid the foundation for broader regional integration. How can those help water diplomats in their negotiation and decision-making process?

3. What diplomatic pathways do the major pollution incidents along the Rhine River provide for water professionals in their solution to transboundary water problems?

4. What were the initial efforts to address the Rhine River pollution culturally, and why were they often inadequate?

5. What were the key components of the Rhine Action Programme (RAP) launched in 1987, and how did they contribute to the the river's recovery?

6. What is the starting point for water diplomats to tackle international water pollution? Is it the large number of ships navigating on the Rhine which have been

discharging oil, garbage, and other pollutants into the river since very ancient times?

The Rhine River's pollution history is deeply intertwined with Europe's industrialization, particularly during the 19th and 20th centuries. As the Industrial Revolution took hold, the river became a conduit for waste disposal from burgeoning industries, urban centers, and agricultural activities. By the mid-20th century, the Rhine was often described as one of the most polluted rivers in Europe.

Chemical industries along the riverbanks discharged hazardous substances, including heavy metals, nitrates, and phosphates. These pollutants led to a dramatic decline in water quality, adversely affecting aquatic life and posing significant health risks to humans. Major pollution incidents, such as the Sandoz chemical spill in 1986 near Basel, Switzerland, which released tons of toxic chemicals into the river, highlighted the urgent need for concerted pollution control.

Initial attempts to address Rhine pollution began in the post-World War II era. Recognizing the transboundary nature of the pollution problem, countries along the Rhine initiated cooperative efforts. In 1950, the International Commission for the Protection of the Rhine (ICPR) was established, marking a significant step towards coordinated action. However, early measures were often inadequate due to limited enforcement capabilities and the ongoing prioritization of industrial growth over environmental protection.

The turning point for the Rhine's recovery came in the latter half of the 20th century with the implementation of more rigorous and scientifically informed pollution control strategies. The following decades saw the introduction of comprehensive measures aimed at reducing industrial discharges, improving wastewater treatment, and restoring ecological health.

One of the most significant initiatives was the Rhine Action Programme (RAP) launched in 1987. The RAP aimed to drastically reduce pollution levels by 2000 through a series of targeted actions:

Industrial Effluent Control: Stringent regulations were imposed on industries, requiring the implementation of advanced treatment technologies to reduce the discharge of hazardous substances. Regular monitoring and reporting ensured compliance with these standards.

Urban Wastewater Treatment: Investment in wastewater treatment infrastructure was significantly increased. The construction and upgrading of sewage treatment plants across the Rhine Basin played a crucial role in reducing the input of organic pollutants and nutrients.

Agricultural Practices: Efforts were made to promote sustainable agricultural practices to reduce runoff of fertilizers and pesticides. Programs encouraged farmers to adopt methods that minimized nutrient leaching and soil erosion.

Accident Prevention and Response: After the Sandoz incident, protocols for accident prevention and emergency response were strengthened. These measures included improved safety standards for chemical storage and transportation, as well as enhanced coordination among riparian countries for swift response to pollution incidents.

The concerted efforts to clean up the Rhine have yielded remarkable results. By the early 2000s, the water quality of the Rhine had significantly improved. Pollutant concentrations dropped dramatically, and biodiversity began to recover. The return of species such as salmon, which are highly sensitive to water quality, marked a significant ecological milestone.

The success of the Rhine's rehabilitation can be attributed to several factors. The cooperative framework provided by the ICPR facilitated consistent and coordinated action among the riparian countries. The adoption of the European Union's Water Framework Directive in 2000 further bolstered efforts by establishing comprehensive water management objectives across EU member states.

The Rhine's journey from a polluted river to a model of recovery offers valuable lessons for water pollution prevention and control worldwide. The key among these lessons is the importance of international cooperation in managing transboundary water bodies. The Rhine experience underscores the need for robust legal frameworks, regular monitoring, and adaptive management practices to address evolving pollution challenges.

Despite the progress made, the Rhine faces ongoing challenges. Climate change, population growth, and emerging pollutants such as pharmaceuticals and microplastics present new threats to water quality. Continued vigilance, innovation in pollution control technologies, and adaptive management strategies will be essential to safeguard the Rhine's health in the future.

The Rhine River's transformation from a heavily polluted waterway to a cleaner and more biodiverse river demonstrates the effectiveness of coordinated pollution control measures. The historical context of industrial pollution, the comprehensive prevention strategies implemented, and the ongoing commitment to environmental stewardship provide a blueprint for addressing water pollution in other regions. As new challenges emerge, the lessons learned from the Rhine will continue to inform and inspire efforts to protect and restore vital water resources globally.

Access to More Working Knowledge for Better Communication

1. Read the multifaceted damage caused by water pollution and discuss in pairs or at a mock negotiating session about how it can affect health, the environment, and the economy.

—Human Health Risks: Contaminated water sources can lead to serious health issues, including gastrointestinal infections, cholera, and other waterborne diseases.

—Ecosystem Disruption: Water pollution harms aquatic ecosystems by disrupting the food chain and harming fish and other wildlife, leading to reduced biodiversity.

—Economic Impact: Polluted water affects industries such as fishing, tourism, and recreation, resulting in economic losses and job reductions.

—Decreased Water Quality: Pollution can degrade the quality of drinking water, necessitating expensive purification processes and increasing the cost of water treatment.

—Soil Contamination: Runoff from polluted water can lead to soil contamination, affecting agriculture and food safety by introducing harmful substances into crops.

—Habitat Destruction: Polluted water bodies can lead to the destruction of critical habitats such as wetlands, which play vital roles in water filtration and flood control.

—Climate Change Vulnerability: Polluted water bodies are often less resilient to climate change impacts, such as droughts and floods, exacerbating environmental degradation.

—Increased Algal Blooms: Nutrient pollution can cause harmful algal blooms, which produce toxins that affect water quality and harm aquatic life.

—Displacement of Communities: Pollution can render water sources unusable, forcing communities to relocate or find alternative water sources, leading to social and economic instability.

—Long-term Health Effects: Chronic exposure to polluted water can lead to long-term health effects, including developmental issues in children and increased cancer risks.

2. Find out who else have been involved in water pollution prevention and control besides the following list of institutions and talk about what they have done in addressing global water issues.

North American Commission for Environmental Cooperation(CEC) 北美环境合作委员会

Southern African Development Community(SADC) 南部非洲发展共同体

Association of Southeast Asian Nations(ASEAN) 东南亚国家联盟

Pacific Islands Forum(PIF) 太平洋岛国论坛

International Organization for Standardization(ISO) 国际标准化组织

International Maritime Organization(IMO) 国际海事组织

World Meteorological Organization(WMO) 世界气象组织

International Association of Water Quality(IAWQ) 国际水质协会

International Network of Basin Organizations(INBO) 国际流域组织网络

International Commission for the Protection of the Danube River(ICPDR) 多瑙河保护国际委员会

International Commission for the Protection of the Rhine(ICPR) 保护莱茵河国际委员会

Great Lakes Commission(GLC) 大湖区委员会

Caribbean Community(CARICOM) 加勒比共同体

Global Environment Facility(GEF) 全球环境基金

United Nations Environment Programme(UNEP) 联合国环境规划署

World Health Organization(WHO) 世界卫生组织

United Nations Educational, Scientific and Cultural Organization (UNESCO) 联合国教育、科学及文化组织

Food and Agriculture Organization (FAO) 联合国粮食及农业组织

World Bank (WB) 世界银行

European Environment Agency (EEA) 欧洲环境署

European Union (EU) 欧盟

Organization for Economic Cooperation and Development (OECD) 经济合作与发展组织

International River Foundation (IRF) 国际河流基金会

United Nations Economic Commission for Europe (UNECE) 联合国欧洲经济委员会

International Fund for Agricultural Development (IFAD) 国际农业发展基金

African Development Bank (AfDB) 非洲开发银行

Further Recommended Reading

[1] Gleick, P. H. *Water in crisis: a guide to the world's freshwater resources*. Oxford: Oxford University Press, 2014.

[2] Saxena, N., Islam, M. M., and Sharma, D. *Water pollution and remediation: a global concern*. Cham: Springer, 2024. https://doi.org/10.1007/978-3-031-76301-4.

[3] World Health Organization. *Guidelines for drinking-water quality: fourth edition incorporating the first and second addenda*. Geneva: World Health Organization, 2022. https://www.who.int/publications/i/item/9789240045064.

Unit 9 Construction of Hydropower Stations

Background Study for Successful Intercultural Communication

Read authentic Chinese and English topical texts in advance before making a glossary for their associated terminology and jargon as follows.

hydropower station 水电站
turbine installation 涡轮安装
reservoir capacity 水库容量
hydrological studies 水文研究
site selection 选址
construction permits 建设许可
civil engineering 土木工程
infrastructure development 基础设施发展
siltation management 淤积管理
grid connection 电网连接
community engagement 社区参与
safety measures 安全措施
project financing 项目融资
stakeholder consultation 利益相关者咨询
operational maintenance 运行维护
transmission lines 输电线路
hydrokinetic energy 水动能
pumped storage 抽水蓄能
economic feasibility 经济可行性
carbon footprint 碳足迹
site survey 工地勘测

water flow assessment 水流评估
energy demand 能源需求
construction timeline 建设时间表
project management 项目管理
construction technology 建筑技术
environmental sustainability 环境可持续性
mechanical systems 机械系统
sediment transport 沉积物运输
local communities 当地社区
public consultation 公众咨询
hydraulic engineering 水力工程
energy production capacity 能源生产能力
biodiversity conservation 生物多样性保护
Gabcikovo-Nagymaros Project (Hungary and Slovakia) 加布奇科沃-大毛罗什项目(匈牙利和斯洛伐克)

Text A
The Gabcikovo-Nagymaros Project on the Danube River

Pre-reading Questions

1. How will water diplomacy facilitate project collaboration, resources sharing, conflict prevention, and dispute settlement when it comes to disagreement over the construction of hydropower stations?

2. The construction of water infrastructure can affect regional water runoff and potentially alter the local landscape. How could those changes influence local cultures and traditions that are deeply connected to the river and its seasonal cycles?

3. What are the potential environmental consequences of constructing hydropower stations on shared rivers for up-stream and down-stream riparians?

4. How does effective stakeholder engagement enhance the decision-making process for transboundary hydropower projects, fostering collaboration and

addressing diverse interests among countries and communities sharing water resources?

5. How can the economic advantages of hydropower generation be effectively balanced with social and environmental priorities to ensure sustainable development and equitable outcomes for all stakeholders?

6. Hydropower infrastructure is deeply interconnected with water management, and hydropower projects constructed on transboundary rivers play a significant role in regulating regional water resources. How do such projects influence the relationships among regions and countries sharing those vital water resources?

7. Joint hydropower projects often require the participation of people from different cultural backgrounds. This promotes European cultural integration and the exchange of technical experiences, contributing to the development of a common water culture. Is the EU experience relevant to water projects in other regions?

The Gabcikovo-Nagymaros Project (GNP) stands as a significant case study in the field of water diplomacy and conflict resolution. The Project is situated on the Danube River, spanning the border between Hungary and Slovakia. This hydroelectric project consists of two main components: the Gabcikovo Dam in Slovakia and the Nagymaros Dam in Hungary. The section of the Danube River involved in the project stretches approximately 200 kilometers, lying between the capital cities of Bratislava in Slovakia and Budapest in Hungary.

The origin of the GNP dates back to the 1950s when Hungary and Czechoslovakia recognized the potential benefits of harnessing the Danube River's resources. In 1977, the two countries signed the Budapest Treaty, agreeing to construct a system of locks, dams, and canals near the towns of Gabcikovo in Czechoslovakia and Nagymaros in Hungary. The project's primary objectives were to improve navigability, control flooding, and generate hydroelectric power, contributing to the economic development of both nations.

The project proceeded with significant momentum until the mid-1980s when environmental concerns began to surface. Environmentalists and scientists in Hungary raised alarms about the potential ecological impacts, including the disruption of natural habitats, reduction in groundwater levels, and negative effects on water quality. The burgeoning environmental movement, coupled with political changes as Hungary transitioned towards a more open system, intensified public opposition to the project.

In 1989, Hungary, citing environmental and economic concerns, unilaterally suspended work on the Nagymaros segment. This decision was met with strong opposition from Czechoslovakia, which had already completed significant construction on the Gabcikovo segment. The suspension led to a diplomatic standoff, with Czechoslovakia accusing Hungary of breaching the Budapest Treaty of 1977.

The conflict escalated to the point where both nations sought resolution through international legal mechanisms. In 1992, the case was brought before the International Court of Justice (ICJ). The court's ruling in 1997 was a complex decision that recognized both nations' legal rights and responsibilities. It affirmed that Hungary had violated the treaty by suspending work but also acknowledged that Czechoslovakia (by then, Slovakia) had overstepped by unilaterally diverting the Danube's flow to complete the Gabcikovo segment.

The ICJ's ruling underscored the necessity of cooperative management of shared water resources, emphasizing that both parties should negotiate a mutually acceptable solution. Despite the ruling, finding common ground proved challenging due to differing national interests and environmental priorities.

In the years following the ICJ's decision, Hungary and Slovakia engaged in protracted negotiations, seeking to balance economic development with environmental preservation. The Gabcikovo segment remains operational, providing hydroelectric power and improved navigation, but environmental concerns persist. The altered hydrological regime has affected local ecosystems, and debates continue over mitigating the ecological impacts.

The Gabcikovo-Nagymaros conflict highlights several critical lessons for transboundary water management. It underscores the importance of incorporating environmental considerations into large-scale water projects, the need for robust legal frameworks to resolve disputes, and the essential role of diplomacy in balancing competing interests. Moreover, it illustrates how political transitions and evolving public attitudes towards environmental conservation can profoundly influence the trajectory of infrastructure projects.

The GNP serves as a poignant example of the complexities inherent in managing shared water resources. While the project achieved some of its initial goals, it also sparked a significant conflict that required legal intervention and ongoing diplomatic efforts. As global water resources face increasing pressures from climate change, population growth, and industrialization, the lessons from the GNP underscore the necessity of cooperative, sustainable, and adaptive approaches to water diplomacy.

Access to More Working Knowledge for Better Communication

1. Read the excerpt below to gain a concise understanding of the process involved in constructing hydropower stations.

The construction of a transboundary hydropower project is a complex and collaborative process that involves several stages. The first phase includes comprehensive feasibility studies, such as site selection, environmental and social impact assessments, and economic evaluations. Given the fact that those projects span multiple countries, regional cooperation is crucial, requiring bilateral or multilateral agreements to address water-sharing, energy production, and environmental protection. During the design and engineering phase, engineers create detailed plans for the dam, power station, transmission lines, and related infrastructure while considering the potential impacts on local communities and ecosystems. Special attention is paid to the concerns of stakeholders in neighboring countries, ensuring that local traditions and practices tied to the river's seasonal changes are respected.

The construction phase focuses on building the required infrastructure, such as dams, reservoirs, spillways, and powerhouses, while managing water flow and diverting rivers to facilitate the process. Cross-border communication and shared responsibilities between the countries are key to minimizing negative impacts on water usage and the environment. Once the project is operational, continuous monitoring of water levels, sediment transport, energy production, and environmental health ensures that the hydropower facility functions as intended while limiting adverse effects on surrounding areas. Hydrological data sharing becomes a vital tool in ensuring transparent management and addressing potential conflicts between the riparian states.

Legal frameworks must be established to ensure compliance with international and national laws governing water use and energy distribution. Those agreements also include mechanisms for dispute resolution, a structured way to address potential conflicts over water resources or environmental degradation. Finally, ongoing maintenance and monitoring of the hydropower infrastructure are necessary for its long-term success with its focus on sustainability, stability, and minimizing inter-country tensions. Therefore, transboundary hydropower projects require extensive international collaboration, careful planning, and constant cooperation to manage the technical, environmental, social, and legal challenges associated with transboundary water governance.

2. Explore the following statistics and statements from 2024 World Hydropower Outlook about hydropower projects and the construction of hydropower stations to gain deeper insight into their development, functionality, and impact.

—Numerous hydropower projects serve multiple functions beyond generating electricity, offering infrastructure to deliver clean water for residential, industrial, and agricultural use, as well as supporting recreation and transportation services.

—The International Energy Agency (IEA) reported that in 2019, hydroelectric power accounted for approximately 16% of global electricity generation, making it the largest renewable energy source for electricity.

—Well-designed and effectively managed hydroelectric power plants have the potential to support biodiversity conservation. This can be achieved through the installation of fish passage systems, which allow fish to migrate freely, and the creation of habitats that can benefit aquatic species, as well as by ensuring that minimum flow levels are maintained, which helps preserve the ecological balance of the water body.

—Global hydropower fleet grows to 1,412 GW in 2023, but five-year rolling average shows downward trend.

—A growth rate of just over 26 GW per year from now to 2030 is needed to stay on track with net zero targets.

—Hydropower is the largest single source of renewable energy, with pumped storage hydropower providing more than 90% of all stored energy in the world.

—It is estimated that around twice the amount of hydropower that is currently installed is needed for net zero scenarios by 2050.

—To double hydropower capacity by 2050, a cumulative investment of approximately US $3.7 trillion is required, or about US $130 billion annually. This equates to more than double the current level of funding.

—IHA's assessment of the "big 100" pipeline of projects under development indicates that this acceleration is within reach for the early years of the next decade, but more action is needed over the longer term.

—Hydropower is among the best ways to mitigate for droughts. IHA estimates that through the water storage function of its reservoirs, the hydropower industry prevents over US $130 billion in annual GDP losses from drought incidents.

Development of Intercultural Communication Skills

1. How can water professionals effectively engage with local communities from diverse cultural backgrounds during the site selection process to ensure that their concerns are considered in the decision-making?

2. How can project managers effectively communicate with local indigenous communities who may have different cultural values, beliefs, and ways of life during the planning phase of a hydropower station construction? If those communities have spiritual attachments to the land where the dam is to be built, how can the construction team convey the benefits of the project while respecting their cultural sensitivities?

3. In a multi-ethnic region where a hydropower station is being constructed, different ethnic groups may have different communication styles. Some may be more direct, while others may be more indirect. How can the construction company adapt its communication strategies to ensure that all stakeholders, regardless of their ethnic backgrounds, understand the project's goals, progress, and potential impacts?

4. Different cultures may have varying perceptions of what "development" means. How can international development agencies and construction companies

work with local communities to align the concept of development brought about by the hydropower station with the local cultural understanding? For instance, some cultures may prioritize environmental harmony and community-based livelihoods over economic growth alone.

5. In some cultures, there may be a strong sense of community-based decision-making, while in others, top-down decision-making is more common. How can the construction of a hydropower station navigate those differences in decision-making cultures when it comes to land acquisition, project implementation, and benefit-sharing arrangements?

6. In regions where multiple languages are spoken, what are the best practices for providing accurate and understandable information about the hydropower station construction to all concerned parties? This includes safety information, construction timelines, and environmental impact assessments. How can language barriers be overcome to ensure effective communication?

7. When communicating technical details about the hydropower station construction to non-technical local communities, how can the use of jargon be minimized, and how can complex concepts be translated into simple, culturally-appropriate language?

8. After the construction of the hydropower station, how can the operating company maintain positive intercultural relations with the local communities? This may involve dealing with issues such as employment opportunities, cultural integration of the company's staff with the local population, and community-driven development initiatives.

9. In the cultures where there is a strong sense of reciprocity, how can the hydropower station operator show gratitude and contribute to the local community in a way that is culturally meaningful? This could include supporting local cultural events, educational programs, or infrastructure development that aligns with the community's values.

Text B
Hydropower Stations in the Lancang-Mekong River Basin

Pre-reading Questions

1. Intercultural communication among the countries in the Lancang-Mekong River Basin is rich and diverse, playing a vital role in promoting regional cooperation and development. Would you please provide more details?

2. In the process of water diplomacy for hydropower, representatives from different countries communicate and interact, which promotes mutual understanding of each other's cultures, values, and policies. What can water diplomats do today to help themselves break down cultural barriers in their upcoming work?

3. What factors prompted the construction of hydropower stations along the Lancang-Mekong River?

4. Can you identify some of the major hydropower projects on the upper Mekong (Lancang) River, and what are their main goals?

5. What concerns have the downstream countries raised regarding China's hydropower development in the region?

6. How do both bilateral and multilateral dialogues play a role in promoting sustainable water resource management within the basin?

The Lancang-Mekong River, known as the Lancang River in China and the Mekong River downstream, has a diverse array of ecosystems and supports agriculture, fisheries, and livelihoods across its basin. China's early interest in harnessing the river's hydropower potential began in the 1980s, with the construction of the Manwan Dam in Yunnan Province. This marked the beginning of extensive hydropower development on the upper reaches of the river.

China's strategic emphasis on hydropower led to the construction of several large dams, including the Dachaoshan, Jinghong, Xiaowan, and Nuozhadu dams. Those projects aimed to meet the growing energy demands of China's southwestern provinces and to foster economic development. However, the rapid pace of dam construction raised concerns among downstream countries about the potential impacts on water flow, sediment transport, and ecological health.

The development of hydropower stations along the Lancang-Mekong has been a source of disputes among the riparian states. Downstream countries, particularly Thailand, Cambodia, and Vietnam, have expressed concerns over the alteration of natural flow regimes, which affects agriculture, fisheries, and local livelihoods. China, Myanmar, Laos, Thailand, Cambodia, and Vietnam all possess varying degrees of hydropower resources and have been actively developing them, albeit in different ways and at different scales. Virtually every Mekong country faces objections—whether from neighbours, civil society or international NGOs—over specific hydropower schemes. The Xayaburi and Don Sahong mainstream hydropower stations in Laos, for instance, have drawn strong objections from Vietnam and Cambodia, which fear damage to fisheries, sediment loss and altered flow regimes downstream. The Mekong Delta, a critical region for rice production in Vietnam, has been particularly vulnerable to changes in water and sediment flow caused by upstream dams.

The Tonle Sap Lake in Cambodia which relies on the Mekong's seasonal flood pulse has experienced fluctuations in water levels that threaten its unique ecosystem and the livelihoods of communities dependent on fishing. Those environmental and socio-economic impacts have fueled tensions and criticism towards hydropower development schemes of upstream riparians.

Moreover, the absence of a comprehensive and legally binding framework for transboundary water management in the basin has complicated the disagreement. While the Mekong River Commission (MRC), established in 1995 by Thailand, Laos, Cambodia, and Vietnam, provides a platform for cooperation, China and Myanmar are not full members, limiting the effectiveness of basin-wide governance.

Despite the disputes, there have been significant cooperative efforts to manage the impacts of hydropower development in the Lancang-Mekong River Basin. The need for collaboration became increasingly evident as the adverse effects of dam construction began to manifest. The establishment of the Lancang-Mekong Cooperation (LMC) framework in 2016 marked a crucial step towards regional dialogue and cooperation.

The LMC, comprising all six riparian countries, aims to promote sustainable development and manage shared water resources effectively. Under this framework, China has engaged more actively with downstream countries, sharing hydrological data and coordinating water releases during critical periods. This data-sharing initiative has been instrumental in improving flood and drought management, thus mitigating some of the negative impacts of upstream dam operations.

Additionally, bilateral and multilateral dialogues have facilitated technical and financial support for sustainable water resource management. For instance, the Mekong-Lancang Cooperation Special Fund, established by China, finances projects that address water resource challenges and promote sustainable development in the basin.

Negotiation has played a pivotal role in managing the disputes and fostering cooperation in the Lancang-Mekong River Basin. Multilateral platforms like the MRC and LMC provide essential venues for dialogue, allowing the countries to articulate their concerns and negotiate mutually acceptable solutions. For instance, the LMC's Joint Working Group on Water Resources has facilitated regular consultations and joint studies on the impacts of hydropower projects.

However, challenges remain in achieving a comprehensive and legally binding agreement that ensures equitable and sustainable water resource management. The diverse interests and priorities of the riparian states, coupled with power asymmetries, complicate negotiation processes.

It is advisable for future negotiations to focus on developing a robust and inclusive framework that incorporates the interests of all riparian countries. Strengthening the role of the MRC and encouraging China and Myanmar to become full members could enhance the effectiveness of transboundary water governance. Additionally, fostering greater transparency and trust through continuous data sharing and joint monitoring initiatives will be crucial.

The construction of hydropower stations in the Lancang-Mekong River Basin exemplifies the intricate balance among dispute, cooperation, and negotiation in transboundary water management. While the development of hydropower has brought economic benefits, it has also generated significant environmental and socio-economic challenges for both upstream and downstream countries. The evolution of cooperative frameworks such as the LMC and ongoing negotiation efforts highlight the potential for collaborative solutions.

Access to More Working Knowledge for Better Communication

1. Investigate other entities that have been involved in the construction of hydropower projects beyond the following list of companies and corporations and discuss their contributions and achievements in a group.

China Three Gorges Corporation (CTG)

SINOHYDRO

Power Construction Corporation of China (POWERCHINA)

ANDRITZ

Voith Hydro

Alstom (GE Hydro Solutions)

Bechtel Corporation

Salini Impregilo (now Webuild Group)

STRABAG SE

Astaris (Merged with Webuild Group)

ACCIONA

Lahmeyer International (now Tractebel Engineering GmbH.)

SNC-Lavalin (now AtkinsRéalis)

Punj Lloyd Group

Enel Green Power

2. Consider the findings of the 2019 MRC Council Report, reflect on the challenges that future water professionals may face, and set the priorities.

The construction of 11 large hydropower dams along the Mekong River's lower mainstream, coupled with the 120 tributary dams planned by 2040, poses a serious threat to the region's ecological integrity, economic stability, and food security.

Those hydropower projects are expected to reduce the amount of sediment reaching the Mekong Delta by an alarming 97%, a critical issue since sediment plays an essential role in enriching the basin, sustaining agriculture, supporting fisheries, and maintaining water quality—key components of the region's economy.

The impacts of mainstream dam reservoirs, flood-control infrastructure, and barriers to fish migration further jeopardize the ecosystem, introducing harmful disruptions to the natural rhythms of the river. According to the Mekong River Commission (MRC), those changes will lead to a dramatic decline in fish stocks, with total fishery biomass projected to reduce by 35%~40% by 2020 and escalate to a 40% ~ 80% decline by 2040. The losses are expected to hit individual countries disproportionately, with predicted reductions in fish catches of 55% for Thailand, 50% for Laos, 35% for Cambodia, and 30% for Vietnam.

Moreover, the construction of dams will severely impact migratory fish species, making it impossible for them to survive in the reservoirs of hydropower projects planned for 2020 and 2040. This elimination of migratory fish will have cascading effects on the river's biodiversity and the livelihoods dependent on fisheries. Compounding those challenges, climate change may amplify the adverse effects, with the combined loss of fish stocks and environmental degradation predicted to lead to "acute levels of food insecurity" in vulnerable communities, particularly in Laos and Cambodia.

Those findings highlight the need for comprehensive planning, collaborative governance, and sustainable alternatives to mitigate the socio-economic and environmental impacts of hydropower development in the Mekong region.

Further Recommended Reading

[1] International Hydropower Association. *Hydropower sustainability assessment protocol*. London: International Hydropower Association, 2020.

[2] International Hydropower Association. *2024 world hydropower outlook*. London: International Hydropower Association, 2024. https://www.hydropower.org/publications/2024-world-hydropower-outlook.

[3] Mekong River Commission. *The MRC Council Study and the impacts of hydropower development on the Mekong*. Bangkok: Mekong River Commission, 2019. https://www.internationalrivers.org/wp-content/uploads/sites/86/2021/03/tragic-trade-offs-the-mrc-council-study-and-the-impacts-of-hydropower-development-on-the-mekong.pdf.

Unit 10 People-to-people Water Diplomacy

Background Study for Successful Intercultural Communication

Read authentic Chinese and English topical texts in advance before making a glossary for their associated terminology and jargon as follows.

cultural exchange 文化交流
people-to-people exchange 人与人交流
mutual understanding 相互理解
cross-cultural dialogue 跨文化对话
non-governmental interaction 非政府互动
community building 社区建设
trust-building 信任构建
educational exchange 教育交流
civic engagement 公民参与
social integration 社会融合
peacebuilding initiatives 和平建设倡议
humanitarian engagement 人道主义参与
collaborative network 协作网络
social capital 社会资本
dialogue platforms 对话平台
international solidarity 国际团结
NGO partnerships 非政府组织合作伙伴关系
peace corps initiatives 和平队倡议
volunteer diplomacy 志愿者外交
shared values 共同价值观
global citizenship 全球公民意识

bilateral engagement 双边接触
cross-sector collaboration 跨部门合作
trust networks 信任网络
community diplomacy 社区外交
knowledge sharing 知识共享
cultural heritage preservation 文化遗产保护
interfaith dialogue 跨宗教对话
social cohesion 社会凝聚力
climate diplomacy 气候外交
sustainable partnership 可持续伙伴关系
inclusive diplomacy 包容性外交
participatory governance 参与式治理
educational outreach 教育推广
multicultural collaboration 多元文化合作
shared prosperity 共同繁荣
cultural co-creation 文化共创
intergenerational dialogue 代际对话
community resilience 社区韧性
dialogue facilitation 对话促进
cross-border networks 跨境网络
peace mediation 和平调解
cultural immersion 文化融入
shared responsibility 共同责任
multilateral cooperation 多边合作
conflict resolution 冲突解决
interpersonal trust 人际信任
cultural resilience 文化韧性
network building 网络构建
bridge-building 桥梁搭建
joint initiatives 联合倡议
cultural innovation 文化创新
youth engagement 青年参与
women in diplomacy 女性在外交中的作用

volunteer initiatives 志愿者倡议
cross-community dialogue 跨社区对话
social partnerships 社会伙伴关系
cultural bridge 文化桥梁
dialogue circles 对话圈

Text A
Roles of Individuals in Water Conflict and Cooperation

Pre-reading Questions

1. "Attentively listen to the statements and demands of all parties involved to understand their core concerns and interests regarding transboundary water issues" suggested an experienced water diplomat. What do you think of his or her suggestion?

2. "Demonstrate respect for the positions and views of other parties through listening, which helps create a good negotiation atmosphere." Are suggestions like this applicable in interpersonal communication?

3. What should we do to improve our ability to clearly and accurately express our own position and stance on transboundary water issues in order to be successful in water diplomacy?

4. How shall we learn to use simple language by avoiding overly technical or ambiguous language to ensure that all parties in future water diplomacy can understand our meaning?

5. How might the involvement of individuals with technical expertise, like engineers or water specialists, promote cooperation in water management?

6. How can individuals, like Aaron T. Wolf or Anupam Mishra, shape water diplomacy through their work?

7. What do you think are the potential risks of individuals' actions in transboundary water governance, and how might they escalate conflicts or inequalities?

8. What are the benefits of fostering international collaboration in water

management through platforms that allow for the exchange of best practices?

In many instances, key individuals have acted as catalysts, bridging divides among conflicting parties and paving the way for mutual agreements. One notable case is the role of Aaron T. Wolf, a scholar and mediator whose expertise in water conflict resolution has significantly influenced diplomatic approaches to shared water bodies. Through dialogue facilitation and conflict mediation, Wolf has worked on numerous occasions to bring entities who share water resources to the negotiating table, emphasizing shared benefits over competing claims. His approach highlights how academic expertise and diplomatic skills can encourage cooperative management in tense situations.

Similarly, Anupam Mishra, a water conservationist from India, demonstrated the impact individuals can have at the grassroots level. Mishra's advocacy for traditional water management systems inspired collaborative projects across regions sharing water resources. His efforts exemplify how individuals can mobilize local communities and champion sustainable practices that benefit broader transboundary ecosystems.

Beyond negotiators and scientists, activists also play a crucial role in bringing attention to transboundary water issues, particularly when marginalized communities are affected. Maude Barlow, a Canadian water activist, has been a vocal advocate for recognizing water as a human right. Her efforts have extended to highlighting the impacts of large-scale water infrastructure projects, such as dams and diversions, on downstream communities. For example, Barlow's campaigns have influenced public opinion and policy discussions around projects like the Colorado River Compact and agreements involving Great Lakes water resources.

In Southeast Asia, Nguyen Phuong, a Vietnamese environmentalist, has raised awareness about the impacts of dam construction along the Mekong River. By documenting changes in fisheries, sedimentation, and local livelihoods, Nguyen Phuong has given a voice to vulnerable populations downstream of large hydropower projects, ensuring that their concerns are represented in regional dialogues. Such advocacy work demonstrates how individuals can counterbalance power asymmetries and advocate for equitable resource sharing.

In addition to advocacy, individuals often introduce technological and managerial innovations that transform how transboundary water resources are managed. For example, Shimon Tal, a former Israeli water commissioner, was instrumental in implementing advanced water recycling and desalination technologies. These innovations enabled Israel to share water resources effectively with neighboring states, such as Jordan and Palestine, under agreements like the Israel-Jordan Peace Treaty of 1994. Tal's contributions show how individuals with technical expertise can foster sustainability and regional stability.

In another instance, Sundeep Waslekar, founder of the Strategic Foresight Group, spearheaded the concept of "Water Cooperation Quotient" to measure collaborative efforts among countries sharing water resources. His work has influenced multilateral discussions and emphasized how trust-building and data-sharing mechanisms can mitigate water-related conflicts.

While individuals can be powerful agents of change, their actions may also reinforce existing inequalities or escalate conflicts. In the construction of Ethiopia's Grand Ethiopian Renaissance Dam (GERD), key government figures have championed the project as a symbol of national pride and development. However, the leadership's inability to reach early consensus with Egypt and Sudan has led to protracted tensions and accusations of unilateralism. This example demonstrates the dual-edged nature of individual influence, where personal or national ambitions can complicate cooperative efforts.

Similarly, when individual interests align too closely with powerful entities, they risk marginalizing vulnerable populations. Corporate stakeholders, lobbyists, and technocrats who prioritize economic benefits over environmental and social concerns can exacerbate inequities in transboundary water governance.

The cases discussed reveal several critical lessons. First, individuals often serve as catalysts for initiating dialogue, advocating for marginalized voices, and promoting technological innovations. Second, their influence is shaped by broader socio-political contexts, meaning that their actions can lead to either conflict or cooperation depending on the values and strategies they employ. Lastly, ensuring inclusivity and accountability in transboundary water governance requires recognizing and empowering diverse individual actors—from scientists and negotiators to activists and local leaders.

Looking ahead, the role of individuals in transboundary water events will likely expand, particularly as global water challenges intensify due to climate change, population growth, and urbanization. Supporting individuals through capacity-building programs, fostering inclusive platforms for participation, and encouraging transparency in governance can create environments where they can contribute meaningfully to sustainable and equitable water management.

Ultimately, while institutions provide the framework for managing transboundary water resources, it is the individuals who bring humanity, creativity, and vision to these shared challenges, ensuring that cooperation can triumph over conflict.

Access to More Working Knowledge for Better Communication

Review the ideas and perceptions below that people may have about actors in transboundary waters. Then, discuss in pairs or participate in a mock negotiating session to explore and clarify the true roles of key actors in transboundary water management.

—Water diplomats primarily serve national political agendas rather than promote genuine cooperation.

—Transboundary water diplomacy efforts are only undertaken by national governments, excluding other actors.

—Transboundary water diplomats hold complete control over decision-making processes and can make independent decisions.

—Transboundary water diplomats are expected to be experts in all areas involved, from science to law to politics.

—Conflicts in transboundary resource management are swiftly resolved by diplomats through quick interventions.

—Water diplomats are driven purely by idealistic and altruistic motives with no influence of national or organizational interests.

—Successful agreements in transboundary water diplomacy ensure that no future disputes or tensions will arise.

—Water diplomats have immediate and easy access to resources, funding, and support needed to address complex transboundary issues.

—Local community's perspectives and needs are secondary to national political objectives in transboundary negotiations.

—Water diplomats focus exclusively on negotiations and conflict resolution, without needing to address technical and scientific complexities.

—Transboundary water diplomats always maintain complete neutrality and do not bring personal biases into their work.

—Once international treaties and agreements are signed, they are consistently followed without the risk of violations or conflicts.

—Water diplomats are consistently successful in avoiding conflicts and maintaining peace between nations or regions.

—The work of transboundary diplomats is reactive, aimed solely at resolving conflicts rather than proactively building cooperative frameworks.

—The outcomes of transboundary diplomacy efforts are immediate, easily measurable, and visibly impactful from the start.

Development of Intercultural Communication Skills

1. Different cultures have diverse attitudes and beliefs towards water. In some arid regions in China, water is regarded as extremely precious which is agreed upon by some people from other regions, yet the same group of peole may be wastful of water in their daily lives. How would you approach that group of people and educate them about the preciousness of freshwater?

2. In the complex and far-reaching field of water diplomacy, the adage "the small is beautiful" holds profound significance. Future water diplomats, who will be at the forefront of resolving transboundary water issues, forging international cooperation on water resources, and safeguarding global water security, can initiate their journey by focusing on small-scale water-saving actions. What steps can be taken to learn to start from small?

3. Spare some time to pracrtice the diplomatic and respectful tone and avoid using forceful or offensive language. Pay special attention to proposing solutions and suggestions in a mild and constructive way. Besides, what else can be done to improve diplomatic language proficiency?

4. Intercutural communication skills are practical such as "Be aware that some gestures may have different meanings in different cultures", "In some cultures, nodding may mean understanding rather than agreement", "Make appropriate eye contact and maintain a friendly facial expression", "In some cultures, direct eye contact is a sign of honesty and respect, while in others, it should be more moderate". The skills mentioned above are non-verbal and hence often neglected in the negotiating room. How shall we incorporate those skills into our water diplomatic training?

5. Compromise and consensus building are essential in water negotiations and discussions. How shall we prepare ourselves to make reasonable compromises and work towards a consensus in our future water problem solving process?

6. Are there any other specific ways to show respect for cultural differences besides respecting the cultural traditions and taboos of other countries, avoiding imposing our own cultural values and finding alternative solutions to culturally unacceptable water issues?

7. Do you think it is essential to adjust communication methods and strategies according to different cultural backgrounds? Do you believe what works in one culture may not work in another? If the answer is affirmative, be ready to adapt.

8. What are the best practices for navigating cultural sensitivities when addressing water disputes or tensions between communities or countries with different cultural norms and political interests?

9. How can cultural differences be turned into assets when designing joint education or outreach programs to raise awareness about sustainable water practices in transboundary settings?

Text B
People-to-People Exchanges in Water Diplomacy

Pre-reading Questions

1. When negotiating a water-sharing treaty, the use of legal jargon and specific numerical allocations of water resources is common. In people-to-people diplomacy, the communication is more organic, using everyday language. What

kind of language do you think water diplomats attending a community-led water-conservation event in a neighboring country should employ? Here, the conversations were about shared concerns for water quality, not about legal obligations, which helped in creating a positive sentiment that could later influence formal negotiations. What is people-to-people diplomacy? And how does it differ from formal diplomatic negotiations in the context of water governance?

2. In what ways can people-to-people diplomacy help address tensions related to transboundary water issues?

3. How do grassroots engagements, like cross-border visits by farmers or fishmen, humanize different perspectives and build trust?

4. Why is including local voices critical in transboundary water governance, and how can people-to-people diplomacy provide platforms for such inclusion?

5. How can shared cultural practices, stories, and traditions strengthen transboundary water cooperation between nations?

6. What is the importance of academic partnerships, such as the UNESCO-IHP initiatives, in fostering mutual understanding of transboundary water challenges?

7. How can storytelling, documentaries, and citizen-led science projects amplify local voices and promote empathy and cooperation in transboundary water management?

While traditional diplomacy often requires formal training and official roles, water diplomacy presents unique opportunities for individuals without typical diplomatic backgrounds to make significant contributions. Those individuals, acting as "water diplomats", play crucial roles in fostering mutual understanding and mitigating conflicts. Human exchanges of information are essential in this context, as they help build trust and break down stereotypes that often persist in formal diplomatic channels.

Initiatives like cross-border visits involving local farmers and fishermen from both upstream and downstream communities provide platforms for sharing experiences, challenges, and solutions. These grassroots interactions humanize differing viewpoints, creating a foundation for cooperation that complements official negotiations. The "Mekong Water Dialogues" organized by the International Union for Conservation of Nature (IUCN) exemplify this approach by bringing together communities from the Mekong Basin to discuss shared water-related challenges. This grassroots collaboration fosters empathy and highlights the common struggles faced by individuals, regardless of nationality, in transboundary water governance, such as climate change, pollution, and ecosystem degradation.

In Africa's Nile Basin Initiative (NBI), civil society organizations have played a vital role in connecting communities from upstream and downstream countries. Workshops and joint projects focusing on sustainable water use empower local people to directly influence water management discussions. Similarly, in areas with transboundary water issues, local residents have spontaneously initiated exchanges and cooperation. For instance, despite the complex political relations between Armenia and Turkey, local people from both countries have established friendly relationships and jointly participated in decisions regarding the management of the Arpa River and the Akhuryan Dam, helping to ease water resource conflicts at the local level.

Cultural exchange is another vital aspect of people-to-people diplomacy. While primarily state-driven, the Indus Waters Treaty between India and Pakistan has allowed informal interactions among civil society organizations, academics, and cultural groups. These engagements emphasize shared dependence on the Indus Basin and help maintain communication bridges even during the periods of heightened political tension. Academic partnerships and exchanges also provide opportunities for researchers and students from different countries to collaborate on shared water challenges. Programs under the UNESCO-IHP (International Hydrological Programme) encourage young scientists from transboundary basins, such as the Danube River and the Senegal River, to engage in collaborative research, contributing to mutual understanding of shared challenges.

Experts and scholars who are not professional diplomats also play important roles in water diplomacy. Shafiqul Islam, a professor at Tufts University and founding director of the Water Diplomacy Program, has conducted in-depth research on water issues and published numerous works. His contributions provide theoretical support and solutions for resolving water disputes and actively promote water diplomacy. Similarly, Benjamin Pohl, a senior project manager of the policy think tank adelphi, introduced the study "The Rise of Hydro-Diplomacy" at the

World Water Week 2014 in Stockholm. His report emphasizes the need for foreign policymakers and diplomats to embrace the opportunities that water diplomacy can offer and highlights ways to strengthen the diplomatic track of trans-boundary cooperation on water, such as investing more in training and capacity-building.

Murray Clamen, an engineer and retired secretary of the International Joint Commission (IJC), and Daniel Macfarlane, an associate professor at Western Michigan University's Environmental and Sustainability Institute, co-authored the book *The International Joint Commission's First Century*. Their work raises awareness of the IJC, an institution established by the United States and Canada in 1909 to protect the waters along the two countries' border, and promotes the resolution of related issues.

Ferdinand de Lesseps, a French diplomat and constructor of the Suez Canal, exemplifies how individuals can drive significant water-related projects. In 1832, he developed a strong interest in building the Suez Canal while serving as a diplomat in Alexandria. After obtaining the canal construction contract from the Egyptian governor Said in 1854, he organized the Canal Company in 1858 and initiated the project in 1859. The canal was officially opened to traffic on November 17, 1869.

Grassroots organizations and NGOs often play a central role in mobilizing communities to work together. Networks such as EcoPeace Middle East, which includes Israelis, Palestinians, and Jordanians, aim to build regional cooperation by engaging local communities in shared water projects. EcoPeace's Good Water Neighbors project has created joint water management initiatives that highlight the interdependence of riparian communities in the Jordan River Basin. The World Bank's Cooperation in International Waters in Africa (CIWA) project also benefits from the efforts of gender and social inclusion specialists. A specialist led the development of the CIWA Gender and Social Inclusion Framework, urging the project to focus on gender equality. As a gender specialist for the Sahel Groundwater Initiative, the same individual has been working to integrate gender and social inclusion issues into all aspects of the initiative, striving to create conditions for the career development of female hydrogeologists and promoting gender equality in transboundary water resource management.

Documentaries, community-led storytelling, and cross-border journalism play significant roles in shaping public perceptions of transboundary water issues. By amplifying local voices, these mediums bring attention to the human dimension of water conflicts and cooperation. Documentaries like *Killing The Mekong Dam by Dam* have highlighted the lives of downstream communities impacted by upstream development projects, promoting empathy and dialogue. Journalists also contribute by covering water issues to increase public awareness and promoting communication and understanding among different parties. *The Water Diplomat*, a monthly news and intelligence magazine, aims to raise awareness of hydro-political challenges and solutions. It reports water-related current affairs, peace, conflict, and diplomacy, acting as a knowledge broker and facilitating constructive dialogue.

Citizen-led scientific monitoring of water quality, biodiversity, and climate impacts creates shared accountability for transboundary rivers by empowering local people to collect and share data. This grassroots knowledge complements technical expertise. In the Zambezi River Basin, communities from multiple countries have worked together to monitor the health of the river through citizen science initiatives.

Individuals like Zhang Yuan, an ordinary water-saving role model, contribute to water resource protection through their actions and advocacy. By installing water-saving appliances and developing good water-use habits, Zhang has significantly reduced his annual water bill and actively participated in water-saving publicity activities in his community and workplace. Although he is not directly involved in the management of transboundary water resources, his efforts raise awareness of water conservation, contributing to water resource protection at a broader level.

Similarly, Jin Liyong, a retired deputy division-level naval officer, resigned from his position as a senior executive in a state-owned enterprise at the age of 57 to engage in the water-saving irrigation business. His company's smart seepage irrigation system has a water utilization rate as high as 95%. They have also independently developed the "rubber powder-based agricultural and forestry seepage irrigation pipe" from waste tires, breaking the foreign monopoly on seepage irrigation pipe technology. His efforts have promoted China's advanced agricultural water-saving irrigation technology, which has been applied in many domestic projects and is expected to be promoted to more countries and regions in the Middle East and beyond. This contributes to improving global water resource utilization efficiency and alleviating the pressure on transboundary water resources.

People-to-people diplomacy is a vital component of transboundary water governance. By fostering trust, inclusivity, and mutual respect, it complements formal institutional efforts and creates pathways for more sustainable and equitable management of shared water resources. Empowering communities and non-state actors strengthens the resilience of transboundary cooperation, turning shared waters from potential sources of conflict into bridges of collaboration. As global water challenges intensify, the importance of fostering connections among individuals and communities will only grow, ensuring that the shared rivers become symbols of unity rather than division.

Access to More Working Knowledge for Better Communication

1. Read the excerpts on water diplomacy provided below and discuss in pairs to explore the capacities and limitations of potential water diplomats.

—The range of actors involved in water diplomacy is diverse, and the presence of existing or potential conflicts between countries does not inherently justify the need for water diplomacy.

—The relationship between water diplomacy and water cooperation is inherently reciprocal: effective water diplomacy depends on cooperation, while political actions are essential to foster and sustain that cooperation.

—Water diplomacy is poised to play a progressively significant role in preventing, mitigating, and resolving both current and future water conflicts.

—Experience demonstrates that preventing and resolving water-related conflicts primarily rely on processes such as research, fact-finding, negotiation, mediation, and conciliation, all grounded in a comprehensive understanding of social, cultural, economic, environmental, and political contexts.

—While the concept of water diplomacy is interpreted differently by various scholars and organizations, there is a shared recognition of the need to integrate multiple dimensions and align stakeholders' interests throughout the cooperation process.

—Informal diplomacy does not operate in isolation; elevating water diplomacy on the foreign affairs agenda must consider the entrenched political and economic interests tied to river development.

—On September 13, 2024, United Nations Secretary-General António Guterres appointed H. E. Retno L. P. Marsudi from Indonesia as his Special Envoy on Water. This appointment, which was highly anticipated, signifies the creation of a dedicated ambassadorial role focusing on championing and advocating for water and sanitation issues at a high level. It is anticipated that the Special Envoy will elevate the importance of water on political agendas both within and outside the United Nations. The role is also expected to bring together stakeholders, enhance the UN's efforts related to water, and mobilize actions and financial resources to tackle global water management challenges.

2. Explore who else may have played a role in people-to-people water diplomacy beyond the following list of actors and discuss how their contributions might have influenced intercultural communication at the grassroots level. Reflect on their unique ability to bridge cultural divides, build trust, and foster understanding through informal but impactful means. How have these less conventional actors strengthened cross-border cooperation and expanded the reach of water diplomacy through their interactions, perspectives, and community influences?

—farmers and fishermen

—indigenous peoples

—community leaders

—teachers and educators

—researchers and scholars

—youth and students

—Non-Governmental Organizations (NGOs)

—environmental activists and conservationists

—cultural and religious groups

—women and women's organizations

—youth-led initiatives

—civil society organizations

—private sectors

—technology providers

—artists

—filmmakers

—storytellers

—community mediators

—conflict resolution specialists

—civil society groups that encourage collaboration among riparian communities

—citizen-led alliances

—engineers

3. Read the excerpt about water education and reflect on how we can cultivate more effective future water diplomats, while leveraging water diplomacy to enhance and complement governmental diplomatic efforts.

Water education spreads knowledge about water resources, management practices, and conservation, fostering an informed populace capable of ensuring sustainable water use. It also underpins people-to-people diplomacy by equipping communities with the knowledge and skills for meaningful engagement, while diplomacy contextualizes education within geopolitical and cultural frameworks.

To improve public water literacy, it is crucial to raise awareness of the global water crisis, climate change's impact on water, and local water resource dynamics. Practical training in areas like irrigation efficiency, wastewater treatment, and water quality monitoring empowers individuals to adopt sustainable practices. Teaching treaties and regulations governing water, especially in transboundary contexts, foster the understanding of collaboration and accountability. Exploring water's historical, cultural, and societal significance strengthens community connections, while promoting conservation projects encourages collective action.

Youth involvement is extremely vital for the long-term success of water diplomacy. Integrating water diplomacy into school curricula, particularly in regions with transboundary challenges, equips students with awareness and problem-solving skills. Supporting youth-led initiatives, such as river clean-ups and sustainable water advocacy, nurtures leadership and environmental stewardship. Facilitating cross-cultural exchanges and joint projects fosters cooperation and builds lasting relationships among future leaders, laying the foundation for effective transboundary diplomacy.

Combining education and participation ensures the general public and younger generations play active roles in addressing water challenges. Together, they contribute to sustainable and collaborative solutions that strengthen global water diplomacy efforts.

Further Recommended Reading

[1] Keith Roberts. *Fluid futures: integrating cultural values into water policy and governance*. 2024. https://jointactionforwater.org/fluid-futures-integrating-cultural-values-into-water-policy-and-governance/.

[2] Neto, S., and Camkin, J. Water diplomacy: more than a drop of wisdom. *WWP World Water Policy*, 2024, 10(4): 1017–1022. https://doi.org/10.1002/wwp2.12237.

[3] Huntjens, P., Yasuda, Y., Swain, A., et al. *The multi-track water diplomacy framework: a legal and political economy analysis for advancing cooperation over shared waters*. The Hague Institute for Global Justice, 2016. https://siwi.org/wp-content/uploads/2018/01/THIGJ_The-Multi-track-Water-Diplomacy-Framework_Webversion-1-1.pdf.

[4] Naho Mirumachi. Informal water diplomacy and power: a case of seeking water security in the Mekong River Basin. *Environmental Science & Policy*, 2020, 114: 86–95. https://doi.org/10.1016/j.envsci.2020.07.021.

Part Ⅲ

Cases of Water Diplomacy between Non-riparian States

Water diplomacy not only involves traditional co-riparian states, but also stakeholders outside a river basin. Whole-of-nation diplomacy—a new approach in diplomacy among states in EU for instance, is worth attention. This EU strategy for cooperation in the Indo-Pacific not only focuses on traditional political relationships but also includes partnerships on multiple areas of common interest, such as sustainable and inclusive prosperity and green transition.

In Part Ⅲ, case studies of water diplomacy between non-riparian states are offered for readers and practitioners working on complex water issues, aiming to enlarge the range of measures available to mitigate water risk, increase preparedness and recovery for droughts and floods, and offer more cost-effective solutions.

Unit 11 Water Resources Utilization

Background Study for Successful Intercultural Communication

Read authentic Chinese and English topical texts in advance before making a glossary for their associated terminology and jargon as follows.

capacity building 能力建设
cooperative frameworks 合作框架
Murray-Darling Basin Plan 墨累-达令盆地计划
water trading 水资源交易
adaptive management strategies 适应性管理策略
water conservation programs 节水计划
prolonged drought 长期干旱
water reuse and recycling 水的再利用和回收
sustainable development 可持续发展
effective allocation 有效分配
holistic approach 整体方法
watershed 流域
stakeholders 利益相关者
social, economic, and environmental objectives 社会、经济和环境目标
Rhine River Basin 莱茵河流域
pollution and flooding challenges 污染和洪水挑战
coordinated flood management strategies 协调的洪水管理策略
environmental standards 环境标准
Colorado River Compact 科罗拉多河协议
collaborative efforts 协作努力
transboundary water management 跨界水管理
Mekong River Basin 湄公河流域

significantly impact water flow 显著影响了水流
sustainable development and management 可持续性发展和管理
impacts of upstream activities 上游活动的影响
Nile River Basin 尼罗河流域
Nile Basin Initiative (NBI) 尼罗河流域倡议
source of tension and negotiation 紧张和谈判的根源
Shared Vision Program 共同愿景计划
joint projects 联合项目
adaptive management 适应性管理
agricultural practices 农业实践
environmental conservation 环境保护

Text A
Israel's Pioneering Water Solutions and Transboundary Collaborations

Pre-reading Questions

1. Some cultures might be more traditional in their water treatment and resistant to new techniques. How would you, a would-be water diplomat, introduce Israel's advanced water-saving technologies in agriculture, such as drip irrigation, to countries with different cultural attitudes towards water use?

2. Water diplomacy provides good opportunities for riparian and non-ripairan countries to clearly and effectively communicate their interests and concerns in water utilization such as securing water for drinking, power generation or environmental protection. How do experienced water experts usually find common ground and develop mutually beneficial agreements in a diplomatic setting among non-riparians?

3. People from different cultures approach large-scale infrastructure construction in different ways. Do you think it is necessary to encourage other countries to invest in similar desalination initiatives mentioned below and collaborate with a small but developed country like Israel?

4. Some cultures tend to be more risk-averse while others are more open to new ideas. How would you encourage people with different interests and cultural values to jointly invest in water-saving works instead of competing for limited water resources?

5. Some countries may have concerns about possible unfairness in water-sharing agreements. How would you address their concerns and make mutually acceptable proposals that take into account the cultural values of fairness and justice in transboundary collaboration?

Despite its location in a water-scarce region, and unlike neighboring countries dealing with an alarming water deficiency, Israel has an abundant supply of water. In fact, Israel is the only country where the desert is shrinking thanks to the abundance of water for agriculture. Over past decades, Israel managed to maximize its water resources and created a prosperous environmental sector, much like water-rich countries. How has Israel made such an achievement? Of course, its desalination facilities provide most of Israel's water along with drip irrigation, water recycling and sustainable water conservation policies. And how does the country conduct water diplomacy as a water superpower?

Sharing Water Technology: Israel is advanced in water treatment, desalination, and drip irrigation technologies. It actively shares those with other countries, like collaborating with some American, African and Asian countries to build desalination plants and introduce drip irrigation systems, thus strengthening water resource management and utilization. A $1 billion Carlsbad desalination plant was built in 2015 by Israel Desalination Enterprises (IDE) in San Diego County, California after its governor had sought help from Israel to overcome its drought-inflicted water shortage. In China, Israel has helped build a desalination plant in the coastal city of Tianjin, 200 kilometers from Beijing. The Tianjin plant—the largest in China—uses thermal-based desalination which is designed to imitate nature in producing freshwater by evaporation and condensation. It jointly implements water resource development and utilization projects with other countries. For example, it has carried out cooperation projects on water resource development with some riparian and non-riparian states to enhance water resource efficiency.

Promoting Experience Exchange: The Israeli government often participates in international water conferences and forums, actively promoting its water management experience and technological achievements, and strengthening communication and cooperation with other countries. It also works with international organizations to host some water resource management training courses and seminars to enhance its influence in the international water field.

Advocating for Cooperation Mechanisms: Through some platforms, Israel advocates the establishment of reasonable international water resource cooperation mechanisms, aiming to promote cooperation and sharing of water resources among countries and contribute to solving global water resource issues. Israel and the United States have established a close cooperative relationship in the fields of science, technology, environment, food safety, and health. The two countries announced a new five-year joint initiative, planning to invest 70 million US dollars to support climate-smart agriculture and improve the capture, storage, use, and protection of critical water resources in the Middle East and Africa. Israel also has extensive cooperation with the European Union in the field of research and development, including water resource management and clean energy technology. The European Commission has multiple cooperation agreements with Israel, especially in the field of water resources and seawater desalination. The Ministry of Innovation, Science and Technology of the State of Israel and the Department for Science, Innovation and Technology (DSIT) of the United Kingdom of Great Britain and Northern Ireland have a special cooperative relationship, strengthening cooperation in scientific research, technology, and innovation, including sustainability, environmental protection (climate change, clean energy, water resources, etc.), as well as health and medical fields.

Signing Bilateral Agreements: The Israeli government negotiates and signs water resource cooperation agreements with other countries. These agreements cover aspects such as cross-border water distribution, water infrastructure construction, and technology sharing, thus safeguarding its water interests and promoting regional and global water cooperation. Over the years, Israel's geopolitical situation is as heated as its climate. Leaders in the Israeli water industry believe that the country's hydraulic technology could also be extended as an olive branch to its neighbors and non-neighbors. Through diplomatic efforts, groundbreaking water saving technologies have been embraced in some water-scarce countries and regions to tackle water scarcity, with seawater desalination and wastewater reuse emerging as key solutions to limited amount of water drawn from aquifers.

Multilateral Diplomatic Coordination: In multilateral settings, Israel coordinates with other countries to jointly address water issues. It participates in the water resource cooperation dialogues of regional organizations and promotes the conclusion of multilateral water resource cooperation agreements to achieve sustainable utilization of water resources. The Negev Forum's Food Safety and Water Technology Working Group, including the Kingdom of Bahrain, Egypt, Israel, Morocco, the United Arab Emirates, and the United States, promotes the use of existing methods and technologies, as well as the development of new methods to improve food safety and water technology.

Access to More Working Knowledge for Better Communication

1. Read the facts listed below and discuss in pairs or at a mock negotiating session about how water diplomats can help to improve transboundary water utilization with intercultural communication skills.

—Globally, around 70% of freshwater withdrawals are used for agriculture. This includes water for irrigation, livestock, and processing crops. Agriculture uses approximately 2,500 cubic kilometers of water yearly. The share of agricultural water use varies significantly by country and region. Many countries in South Asia, Africa, and Latin America use more than 80% of their water withdrawals for agriculture.

—Industrial processes account for around 20% of global water consumption. The United States and China are among the countries that withdraw the most water for industrial uses. Industrial water is used for a range of applications, including dilution, steam generation, washing, cooling of manufacturing equipment, and energy generation in fossil fuel and nuclear power plants. Industrial water usage tends to dominate in high-income countries and is relatively small in low-income countries.

—Domestic water use, which includes water for drinking, cleaning, washing, and cooking, makes up about 10% of global water withdrawals. China, India, and the United States have the highest domestic water demands, partly due to their large populations, with higher per capita water demands in the United States. However, domestic water use is relatively small compared to agricultural and industrial applications in most countries.

—The global population has surpassed 7.9 billion and is projected to reach 9.7 billion by 2050. This population growth, along with economic development, especially in developing countries, has led to increasing water consumption per capita, putting more pressure on water resources.

—Climate change has introduced severe fluctuations in water availability, with some regions experiencing droughts while others face floods. This has made water management more challenging and necessitates adaptations to ensure sustainable water use.

—As water demand increases due to population growth and economic development, water stress and the risk of water scarcity have become common concerns, particularly in regions with lower water resources and higher population pressures.

—Smart irrigation systems that utilize data and sensors to optimize water delivery can help reduce waste and increase crop yield. Additionally, adopting water-efficient practices and technologies in agriculture can significantly improve water use efficiency.

—Implementing water recycling and reuse systems within industrial processes can drastically reduce freshwater consumption. Advanced filtration technologies and smart water management systems that provide real-time data analytics also contribute to more efficient industrial water use.

—Many households are adopting water-saving habits, such as taking shorter showers and using water-efficient appliances. Technologies like water recycling and desalination can also help diversify water sources and improve domestic water supply.

2. Read the following misconceptions that people have about water in their daily life and dispel them by articulating arguments to foster effective water usage practices in all stakeholders involved.

—Freshwater Is Unlimited: Many people mistakenly believe that freshwater is an inexhaustible resource. However, only about 2.5% of the earth's water is freshwater, and the vast majority of it is trapped in glaciers and polar ice sheets. This leaves less than 1% readily accessible for human use in rivers, lakes, and groundwater.

—Water Scarcity Is Only a Problem in Arid Regions: It is often assumed that water scarcity is confined to arid and drought-prone regions. In reality, water scarcity can occur in any area where water demand exceeds supply, regardless of climate. Factors such as rapid population growth, urbanization, pollution, and inefficient water use can lead to water shortages even in regions traditionally considered water-rich.

—Domestic Use Is the Main Water Consumer: A common misconception is that the water used in homes for drinking, cooking, and bathing constitutes the largest share of freshwater consumption. In fact, agriculture is by far the largest consumer, using about 70% of the world's freshwater primarily for irrigation. Industrial use also represents a significant portion, especially in developed countries.

—Water Conservation at Home Makes the Biggest Difference: While individual efforts to conserve water at home are important, they represent only a small fraction of overall water use. Significant water savings can be achieved through changes in agricultural practices, industrial processes, and water management policies.

—What I Do with My Water Doesn't Make a Difference: Some people believe that their personal water use habits have no significant impact on the bigger picture. However, when multiplied by thousands or millions of users, small changes can lead to substantial reductions in water use. For example, if everyone in the US saved just a few drops of water each day, the results would be staggering.

—Water Management Technologies Are Too Expensive: Although initial costs can be significant, the long-term savings often outweigh these expenses. Technologies such as smart irrigation systems and rainwater harvesting can significantly reduce water bills and conserve resources, leading to financial savings over time. Additionally, many governments and organizations offer incentives and rebates to encourage the adoption of water-efficient technologies.

—Water Management Technologies Are Complicated to Use: Most modern water management systems are designed to be user-friendly and intuitive. With the integration of digital interfaces and automated controls, managing these technologies has become simpler than ever. Users can easily access data through apps and dashboards, providing real-time information on water usage and system performance.

—Bottled Water Is Safer Than Tap Water: The FDA and the EPA share the same exact standards when it comes to the water they are providing. Additionally, bottled water requires much more water and energy to make than simply getting a drink from the tap.

—Fountains Contain Pure Water: There's a romantic notion that bubbling fountains offer a source of pristine and pure water. However, fountains, especially in public areas, often recycle water that has been treated to prevent stagnation but not filtered to drinking standards. This recycled water can harbor contaminants like bacteria or chemicals.

—Everyone Needs to Drink Eight Glasses of Water Per Day: There is little scientific basis for this statement. Actual fluid needs typically depend on body size and energy expenditure (i. e., physical activity). Fluid intake, driven by thirst, allows maintenance of hydration and total body water at normal levels.

—Drinking Water Flushes Toxins from One's Body: If one is not properly hydrated, his or her kidneys don't have the right amount of fluid to remove metabolic wastes as efficiently. In other words, lack of water causes the body to hold in toxins rather than expelling them as required for proper health.

—Individuals with certain health complications, such as high blood pressure and/or swelling of the lower legs (i. e., edema), need to avoid excess water. If one has had a kidney transplant, he or she should also talk with the doctor about their fluid intake.

Text B
UAE's Water Innovation for a Sustainable Future

Pre-reading Questions

1. What qualities should future water diplomats have in order to globally

promote the sharing of water-efficient irrigation techniques and industrial water-recycling methods? Consider how you can navigate cultural differences to ensure that all relevant countries could benefit from these practices.

2. Diplomatic cooperation in water utilization is key to achieving Sustainable Development Goal 6. What would you do to foster similar cooperation among different countries in the future? Think about how you could help to develop more efficient water-sharing mechanisms, invest in water treatment infrastructure, and improve access to clean water in underserved areas.

3. How would you identify the cultural perspectives fueling water disputes and conflicts, work with culturally-aware mediators, and employ culturally appropriate negotiation techniques to reach consensus that respects the norms and interests of every party involved at the diplomatic table?

4. How would you, a water diplomat in training, leverage the opportunity to initiate or participate in cultural exchange programs to enhance your intercultural communication skills? How would you draw the attention of representatives from different countries to each other's culture, water management practices, and the significance of water in their societies, like organizing a water treatment facility tour?

5. Middle Eastern cultures tend to place high value on relationships and consensus-based decision-making, while Western cultures may be more direct and individualistic. How would you tailor your initial outreach to stakeholders from different regions? How would you approach a Middle Eastern water department and a Western water organization before starting a dialogue about a shared water resource project?

The UAE, with its vast deserts and limited natural freshwater sources, has long been a pioneer in finding inventive solutions to the challenges posed by water scarcity. Over the years, the country has invested significantly in both technology and sustainable practices to address not only its domestic water needs but also its goal to become a global leader in water innovation.

One such innovation is the country's investment in desalination technology, which converts seawater into drinkable water. The Middle East produces half of the world's desalinated water, as much as 7.5 million cubic meters per day. But it comes at a cost. The standard thermal desalination process—boiling seawater and capturing and condensing its vapor—is expensive, generating substantial carbon emissions, and producing significant wastewater. While desalination has been a staple solution for water-short nations, the UAE has been integrating renewable energy sources such as solar power into desalination plants. The Mohammed bin Rashid Al Maktoum Solar Park, one of the largest single-site solar parks in the world, has enabled the country to integrate solar-powered desalination, significantly reducing the carbon footprint of this otherwise energy-intensive process.

To further advance desalination technologies, the UAE's partnerships with advanced water tech companies in France, Republic of Korea, and Spain have been established. Leading French water management companies, Veolia and Suez, along with Spanish water company Aqualia, have joined forces with the UAE water institutions to develop and operate top-notch desalination plants. These partnerships leverage their expertise in water treatment and sustainable technologies to address the UAE's unique challenges. Additionally, the Masdar Institute (now part of UAE's Khalifa University) has teamed up with French research institutions to explore innovative materials and processes for desalination, focusing on advanced membrane technologies that enhance efficiency and sustainability. The Korea Water Corporation (K-water) has collaborated with the UAE to jointly explore innovative ways to increase desalination energy efficiency and mitigate environmental impacts. They have contributed valuable expertise in cutting-edge technologies like thermal and membrane desalination systems, essential to supporting the UAE's water demands.

Besides desalination, the UAE is also working on alternative water-saving technologies. One of the more promising innovations is vertical farming, an agricultural method that optimizes water usage by growing crops in stacked layers. This practice dramatically reduces the amount of water needed for irrigation, meanwhile allowing for greater food production in limited spaces. By developing these highly efficient farming systems, the UAE aims to bolster food security while conserving precious water resources.

Another exciting technological breakthrough is the use of cloud seeding. This process, which involves dispersing substances like silver iodide into clouds to stimulate rainfall, has the potential to increase freshwater availability in arid regions. The UAE has positioned itself at the forefront of cloud seeding technology through the efforts of its National Center of Meteorology (NCM). The UAE has also invested heavily in research and development to improve the efficacy of cloud seeding, with a strong focus on developing environmentally safe materials to mitigate potential ecological concerns. Joint research projects formed between China and the UAE target the creation of innovative materials and advanced methods for cloud seeding tailored to the UAE's climatic conditions. These collaborations include studying alternative materials, such as biodegradable particles, to replace traditional substances like silver iodide, ensuring minimal environmental impact. Furthermore, the projects aim to develop specialized techniques suited for the UAE's extreme heat and low humidity, maximizing the potential for successful precipitation enhancement.

To communicate more efficiently with the other Gulf countries, the UAE is actively exchanging knowledge and expertise in cloud seeding technology. This regional partnership facilitates the sharing of best practices and operational data to address common water security challenges across the Arabian Peninsula. By pooling resources and experiences, Gulf countries aim to refine the efficiency of cloud seeding and optimize its implementation on a broader scale. In addition to these cutting-edge innovations, the UAE has prioritized sustainable water management practices such as wastewater treatment and water recycling. By investing in advanced treatment facilities that purify wastewater for reuse in agriculture, landscaping, and even drinking water in some cases, the country is setting an example for the world in responsible water use. These efforts align with the UAE's vision for a circular economy, where water is continuously reused and recycled, contributing to a sustainable water cycle.

The UAE's initiatives not only aim to secure the nation's safe water future but also set up a valuable example for other arid regions struggling with water scarcity. In fact, UAE-based entrepreneurs and institutions are actively collaborating with international organizations, such as the World Economic Forum and the United Nations, to share knowledge, collaborate on research, and scale solutions globally.

The country's leadership in water innovation also goes beyond technology. The UAE has implemented robust policies and legislation that promote water conservation at every level, from agriculture to urban infrastructure. The use of drip irrigation systems, for example, has been widely adopted across the UAE's agricultural sector, ensuring that every drop of water is used efficiently to irrigate crops. The government's long-term vision incorporates sustainable water infrastructure, with projects aimed at storing water in underground reservoirs, reducing evaporation loss, and improving distribution systems.

In addition to technology and policy, the UAE is also focused on educating the public about the importance of water conservation. Campaigns, educational programs, and the integration of water sustainability principles into school curriculums help to instill a culture of water conservation into all generations. These efforts ensure that sustainability is not just a top-down government initiative but a societal norm that every citizen plays a part in upholding.

As the global water crisis escalates, the UAE's forward-thinking approach to innovation and sustainability serves as a model for the world. The combination of groundbreaking technology, efficient resource management, and a strong policy framework, coupled with public awareness and education, has positioned the UAE as a key player in addressing one of the world's most pressing challenges. The country's initiatives offer hope for the future, demonstrating that with the right collaboration and investment, even the harshest climates can be turned into fertile ground for water innovation and sustainability.

Access to More Working Knowledge for Better Communication

1. Read the facts listed below and discuss in pairs about how the water resources is utilized in different major sectors.

—Agriculture is the largest consumer of freshwater, accounting for approximately 70% of total water withdrawals globally.

—The industrial sector follows, using about 20%, while domestic consumption remains around 10%.

—Environmental requirements, though often overlooked, are crucial for ecological balance and need careful management.

—Emerging trends indicate a need for improved efficiency in water use across all sectors.

—Agriculture uses approximately 2,500 cubic kilometers of water yearly. The water is mainly utilized for irrigation, livestock, and processing crops. Several regions face water scarcity due to overconsumption and droughts.

—Industrial processes demand around 1,500 cubic kilometers of water annually. The demand varies greatly among different industries. Manufacturing and energy production typically require more water than services or technology sectors.

—Domestic water use commonly sits at about 600 cubic kilometers per year. This encompasses all household activities such as drinking, cooking, sanitation, and gardening. Urban areas often showcase significantly higher per capita consumption compared to rural settings.

2. Read the following information about water consumption and discuss how would-be water diplomats can use it for advocating for better water policies and practices.

In the early 20th century, global water consumption was relatively low, with only 500 cubic kilometers of water used per year. However, as the population grew and industrialization took hold, water usage surged. By the year 2000, consumption had increased eightfold to 4,000 cubic kilometers per year. This rapid increase put immense pressure on water resources, leading to scarcity in many regions.

Today, approximately 3.6 billion people live in regions with limited water supplies, experiencing water shortages for at least one month each year. By 2050, it is projected that 4.8 to 5.7 billion people will reside in water-scarce areas, primarily due to population growth in emerging nations. The situation is exacerbated by climate change, which alters hydrological cycles, making water more unpredictable and increasing the frequency and intensity of floods and droughts.

Agriculture is the largest consumer of water, accounting for 70% of global water use. Feeding the growing population requires a 50% increase in agricultural production by 2050, which will further strain water resources. In many parts of the world, groundwater supplies are being depleted at alarming rates. For example, the Ogallala Aquifer in the western United States is being used faster than it can be replenished.

Industry also plays a significant role in water consumption, accounting for 20% of global water use. In industrialized nations, industries consume more than half of the water available for human use. The production of biofuels, for example, requires 1,000 to 4,000 liters of water to produce a single liter of biofuel. This has led to increased competition for water resources, particularly in regions where water is already scarce.

Water scarcity is not just a problem for arid regions. Over 40% of the world's population lives in water-scarce areas, and approximately 1/4 of the world's GDP is exposed to this challenge. By 2040, an estimated one in four children will live in areas with extreme water shortages. The fragmentation of water resources, with 276 transboundary basins shared by 148 countries, further complicates the issue. Cooperation is essential to achieve optimal water resources management and development solutions for all riparians.

To address these challenges, countries need to improve the way they manage their water resources. Institutional tools such as legal and regulatory frameworks, water pricing, and incentives are needed to better allocate, regulate, and conserve water resources. Information systems for resource monitoring, decision-making under uncertainty, and hydro-meteorological forecasts are also crucial. Investments in innovative technologies for enhancing productivity, conserving and protecting resources, recycling stormwater and wastewater, and developing non-conventional water sources are essential.

While large-scale changes are necessary, individual actions also play a significant role. Simple steps like fixing leaks, using water-efficient appliances, and reducing water usage in daily activities can make a big difference. For example, using low-flow showerheads and aerators can reduce water usage by up to 50%. In addition, educating others about the importance of water conservation and adopting sustainable practices can help create a culture of water stewardship.

The future of water resources management lies in a combination of technological innovation, policy reform, and individual responsibility. By working together, we can ensure that everyone has access to clean and safe water while preserving this precious resource for future generations. The journey is long and complex, but with determination and cooperation, it is a journey we can succeed in.

Further Recommended Reading

[1] Shuval, Hillel I. *Water quality management under conditions of scarcity: Israel as a case study*. New York: Academic Press, 1980.

[2] Zekri, S. *Water policies in MENA countries*. Switzerland: Springer, 2020.

[3] Lipchin, C., Pallant, E., Saranga, D., and Amster, A. *Integrated water resources management and security in the Middle East*. Netherlands: Springer, 2007.

[4] Amin Shaban. Water resources in the Middle East. In *Satellite monitoring of water resources in the Middle East*, edited by Amin Shaban, 15-30. Cham: Springer International Publishing, 2022.

Unit 12 Water Saving

Background Study for Successful Intercultural Communication

Read authentic Chinese and English topical texts in advance before making a glossary for their associated terminology and jargon as follows.

international stakeholders 国际利益相关者
regional framework 区域框架
sustainable water resources 可持续的水资源
Syrian conflict 叙利亚冲突
semiarid climate 半干旱气候
National Water Carrier Project 国家输水工程项目
water conservation project 节水工程
irrigation water conservation 灌溉节水
urban water conservation 城市节水
industrial water conservation 工业节水
water-saving technology 节水技术
water-saving device 节水器具
water-saving awareness 节水意识
water quantity monitoring 水量监测
water rights system 水权制度
water pricing mechanism 水价机制
water allocation plan 水资源分配方案
river basin management 流域管理
groundwater management 地下水管理
surface water management 地表水管理
desalination project 海水淡化工程
wastewater treatment project 污水处理工程

recycled water utilization 再生水利用
water-saving irrigation project 节水灌溉工程
water conservation strategy 节水战略
water-saving measure 节水措施
water-use efficiency improvement 用水效率提高
rainwater harvesting project 雨水收集工程
water-saving renovation 节水改造
water-saving evaluation 节水评价
water-related infrastructure construction 水利基础设施建设
water management information system 水管理信息系统
water-saving publicity 节水宣传
water conservation laws and regulations 节水法律法规
water-use quota management 用水定额管理
water-saving incentive mechanism 节水激励机制
water-saving demonstration area 节水示范区
water-saving education 节水教育

Text A
Singapore's Water Conservation

Pre-reading Questions

1. How do different cultural attitudes towards water resources, such as views on water as a precious commodity or a common good, impact negotiation positions? How can negotiators bridge the gap when there are significant cultural differences in the perception of water rights and ownership?

2. In water negotiations, how do differences in direct and indirect communication styles among cultures affect the exchange of ideas and the reaching of agreements? How can negotiators adapt their communication styles to better understand and be understood by counterparts from other cultures?

3. What role do cultural values like collectivism or individualism play in water negotiation strategies? How can negotiators respect and incorporate the values and beliefs of different cultures into the negotiation process to build mutual

trust?

4. How does Singaporean culture influence individual and collective water-saving behaviors, and what implications does this have for intercultural communication on sustainable water use?

5. In an intercultural context, how can the cultural norms and values related to water conservation in Singapore be used to inspire and motivate similar behaviors in other countries?

6. What technological innovations has Singapore introduced to save water, and how are these technologies shared and communicated across cultures?

7. How do Singapore's water-saving policies take into account cultural factors, and how can they be effectively communicated and implemented in international cooperation projects?

Water conservation in Singapore is widely recognized as a model of success for cities grappling with water scarcity. In its early years of independence during the 1960s and 1970s, Singapore relied heavily on local catchment water and imports from Malaysia. Recognizing the vulnerability of this arrangement, the government embarked on a mission to diversify and secure its water sources. By the early 2000s, innovations such as NEWater and desalinated water had transformed Singapore into a global leader in water resource management.

Today, the city-state consumes approximately 440 million gallons of water daily, with projections showing that this demand will double by 2065 due to population growth and increased industrial activity. Furthermore, climate change is expected to exacerbate pressure on the country's limited water resources, underlining the critical need for conservation and innovation to ensure long-term resilience.

With no significant aquifers and only a few small rivers, such as the Kallang River and Singapore River which primarily function as urban drainage channels, Singapore faces formidable challenges. Although it receives abundant annual rainfall, its small land area limits the spacc available for water capture and storage. This complex scenario necessitates a groundbreaking strategy—the "Four National Taps".

In 2004, then-Minister for the Environment Lim Swee Say introduced the

"Four National Taps" which have become the backbone of Singapore's water resilience.

The First National Tap is local catchment water. Singapore maximizes rainwater collection through an extensive network of reservoirs, canals, and drains. Today, rainwater is harvested from two-thirds of the nation's land area. The water is funneled into 17 reservoirs, where it undergoes treatment to meet potable standards. Notable milestones include the construction of the Marina, Punggol, and Serangoon reservoirs, which expanded the water catchment area from 50% to 67% of Singapore's surface area. To enhance sustainability, Singapore's dual-system infrastructure separates rainwater and wastewater management. Rainwater is collected and treated for drinking, while wastewater is processed at reclamation plants for reuse. This innovative approach makes Singapore one of the few countries in the world capable of harnessing urban stormwater on a large scale for drinking purposes.

Singapore's Second National Tap is imported water. Under the 1962 Water Agreement, Singapore imports up to 250 million gallons of water daily from the Johor River in Malaysia while providing Johor with treated water equivalent to 2% of the imported volume. As this agreement will expire in 2061, Singapore continues to focus on reducing reliance on imported water through innovation and self-reliance.

NEWater, the Third National Tap, is a cornerstone of Singapore's water strategy, addressing up to 40% of its current water demand. Using advanced processes such as microfiltration, reverse osmosis, and ultraviolet disinfection, reclaimed water is transformed into ultra-clean and high-quality water. NEWater is primarily used for industrial purposes and indirect potable use, as it is blended into raw water reservoirs for further treatment.

The Fourth National Tap, desalination, provides a sustainable and drought-resistant source of freshwater. Singapore currently operates five desalination plants, which meet up to 30% of its water needs. Ongoing research focuses on improving energy efficiency and reducing the environmental impact of desalination technologies.

Public participation is a critical element of Singapore's water management strategy. The National Water Agency, PUB, spearheads efforts to cultivate a water-conscious society. Initiatives such as the annual "Make Every Drop Count" campaign held during Singapore World Water Day in March engage individuals and businesses in water-saving practices. Educational programs, community events, and water audits encourage behavioral change, emphasizing the role of every citizen in securing Singapore's water future.

Singapore's water sustainability achievements are underpinned by extensive global partnerships. Recognizing that challenges transcend national borders, this nation collaborates with other countries to exchange expertise and develop innovative solutions.

Singapore's cooperation with Malaysia extends beyond water imports to include technical exchanges on efficient water utilization and resource management. Through Sino-Singaporean collaboration for industrial wastewater treatment and resource reuse, both nations have advanced membrane technologies for desalination and wastewater recycling. Israel, the United States, and Japan partner with Singapore to innovate water-saving techniques and refine desalination methods. Collaborative projects with Dutch companies focus on water reuse, flood resilience, and adaptation to climate change. Singapore also shares its expertise in rainwater harvesting, urban water management, and waste reduction with ASEAN countries, contributing to regional water security.

As a hub for global water innovation, Singapore hosts the biennial Singapore International Water Week (SIWW). This flagship event convenes experts, policymakers, and industries to explore sustainable water solutions. The resulting partnerships frequently lead to cutting-edge projects that enhance water security worldwide.

Singapore's commitment to sharing knowledge is also exemplified by initiatives such as WaterHub, which trains professionals from developing countries in best practices for water management. These efforts enable global adoption of sustainable techniques and strengthen water resilience in resource-scarce regions.

Access to More Working Knowledge for Better Communication

1. Read the following information on methods that individuals can adopt for saving water and discuss in pairs about how water diplomats can combine these tips with water conservation polices and initiatives at the national and international levels.

Individuals can play a significant role in water conservation through simple yet effective practices at home and in daily life.

Water-Efficient Appliances: Installing water-efficient fixtures such as low-flow toilets and aerated faucets can significantly reduce household water use. For example, upgrading to WaterSense labeled toilets can save up to 13,000 gallons of water annually.

Behavioral Changes: Simple changes in daily habits, such as turning off the tap while brushing teeth or taking shorter showers, can lead to substantial water savings. Using a bucket to collect water during showers can also prevent wastage.

Greywater Reuse: Reusing greywater from sinks, showers, and laundry for toilet flushing or irrigation can reduce overall household water demand. This practice is especially effective in water-scarce regions.

Leak Detection and Repair: Regularly checking for leaks in faucets, toilets, and pipes can prevent significant water loss. Using dye tests to detect invisible leaks is a simple and effective method.

Rainwater Harvesting: Capturing and storing rainwater for irrigation or other non-potable uses can reduce dependence on groundwater and surface water sources.

Smart Water Meters: Installing smart water meters allows individuals to track real-time water usage and identify leaks promptly. This technology provides detailed insights into consumption patterns and helps set conservation goals.

Water-saving initiatives at both the national and individual levels are essential for addressing water scarcity and promoting sustainable water use. By combining regulatory frameworks, technological advancements, and behavioral changes, countries and individuals can make significant progress in water conservation. As water scarcity becomes a growing global challenge, the collective efforts of governments, communities, and individuals will be crucial in ensuring a sustainable water future.

2. Read the following excerpt on water pricing and discuss in pairs about what alternative methods, besides those adopted by Singapore, can be used to promote water conservation.

Water pricing serves as a key economic tool for optimizing water allocation and enhancing efficiency. When implemented effectively, it can become the most powerful mechanism for managing water demand and attaining financial and economic objectives.

By influencing consumer behavior and encouraging efficient use, water pricing can help address water scarcity and promote environmental sustainability. Here is how water pricing works and its impact on water conservation:

Dynamic water pricing adjusts water costs based on supply and demand conditions. For example, in Sydney, Australia, water prices increase when dam storage levels fall below 60% and decrease when they exceed 70%. This approach encourages water consumers to reduce water usage during periods of scarcity, thereby ensuring sufficient water supply without the need for rationing.

Marginal cost pricing involves setting water prices based on the cost of supplying additional units of water. This method incentivizes water conservation by reflecting the true cost of water, including supply costs, opportunity costs, and environmental externalities. For example, studies have shown that marginal pricing can significantly reduce water demand by making consumers more aware of the value of water.

Tiered pricing structures, such as increasing block tariffs, charge higher rates for higher levels of water consumption. This approach encourages water conservation by making excessive water use more expensive. For instance, Beijing has implemented a tiered pricing system that promotes water-saving behaviors among residential and industrial users.

Higher water prices can incentivize investments in water-saving technologies, such as low-flow fixtures and efficient irrigation systems. For example, a study in Bangladesh found that water-saving technologies were most effective when farmers faced marginal pricing rather than fixed fees.

While water pricing can be an effective tool for conservation, it is essential to ensure that it does not disproportionately affect low-income households. Policies like subsidies and targeted pricing can help maintain equitable access to water while still promoting conservation.

Text B
Water-saving Strategies in Denmark

Pre-reading Questions

1. When concerns arise in water negotiations due to cultural misunderstandings, what are the most effective ways to resolve them while respecting cultural diversity? How can negotiators use cultural knowledge to prevent potential disagreement from escalating?

2. How do cultural norms regarding patience and urgency influence the pacing and outcome of water negotiations? Are there cultural-specific negotiation tactics that are effective in water diplomacy, and how can they be integrated into an intercultural context?

3. When talking about Denmark's integrated management strategy with representatives from a country with a fragmented water management system, how would you use intercultural communication skills you learned at school to highlight the advantages of integration and persuade them to consider a similar approach?

4. What are the main water-saving strategies implemented in Denmark, and how do they reflect the country's cultural values and priorities? How does Denmark's approach to water conservation compare with those of other countries, and what can be learned from these differences in an intercultural context?

5. How does Denmark raise public awareness about water-saving, and what role does cultural communication play in promoting these efforts? In an intercultural setting, how can Denmark's experience in educating the public about water conservation be adapted and applied in other countries with different cultural backgrounds?

6. How does Denmark collaborate with other countries on water-saving initiatives, and what challenges and opportunities arise in this intercultural

communication process?

7. What role does Denmark play in sharing its water-saving strategies with developing countries, and how can cultural differences be addressed to ensure successful knowledge transfer?

Denmark is renowned globally for its expertise in water solutions, excelling in water supply, wastewater treatment, and climate adaptation. It offers some of the most influential solutions that benefit the global environment, climate, and public health. The Danish water sector leads Europe in water-related patents and exports per capita, reflecting its innovation and influence. Additionally, Denmark is home to several leading water companies and boasts a robust network of small and medium-sized enterprises, research institutions, and regulatory authorities.

A holistic management strategy defines Denmark's approach to water conservation, integrating groundwater protection, efficient water supply systems, and advanced wastewater treatment. This integrated framework ensures the sustainable management of water resources from source to tap. Danish water utilities implement stringent regulations and monitoring practices to safeguard groundwater sources, ensuring that water abstraction remains sustainable and critical aquifers are preserved.

Advanced technologies form the backbone of Danish water utilities' efforts to minimize water loss and optimize distribution networks. Supervisory Control and Data Acquisition (SCADA) systems are widely used to monitor and control water supply infrastructure, enabling efficient operations and rapid leak detection. Furthermore, innovative solutions such as smart metering and real-time monitoring have been developed to promptly identify and address leaks. These technologies play a critical role in reducing water loss and enhancing the overall efficiency of water distribution.

Public awareness and engagement underpin the success of water conservation strategies in Denmark. Through extensive water-saving campaigns and educational programs, Danish utilities have successfully raised public awareness about the importance of conserving water as a limited resource. These initiatives have resulted in significant reductions in domestic water consumption, with per capita daily water use decreasing by over 20% in the past decade. Additionally, Danish utilities promote the adoption of water-saving technologies and practices, including low-flow fixtures and rainwater harvesting systems, further reducing water consumption at the household level.

Sustainable water management practices have positioned Denmark as a global leader in water conservation. The country actively collaborates with other nations to share its expertise and promote sustainable practices worldwide. Through initiatives such as the Danish Water Partnership (DWP), Denmark works with international organizations, governments, and private companies to facilitate knowledge transfer and the dissemination of advanced water management technologies.

Collaborative efforts of Danish water utilities and research institutions with international partners strengthen global water management capacity. These partnerships focus on Integrated Water Resources Management (IWRM), wastewater treatment, and public water conservation campaigns. Denmark's holistic water management approach and advanced technological expertise are particularly valuable in regions grappling with water scarcity and environmental challenges.

Strong public-private partnerships are a hallmark of the Danish water sector, driving the development and implementation of sustainable water solutions. Danish companies often collaborate with international counterparts to deliver comprehensive water management projects, integrating Danish expertise with local knowledge and resources. These partnerships guarantee that projects are tailored to specific regional needs, maximizing their efficiency and sustainability.

Global initiatives aimed at addressing water challenges benefit greatly from Denmark's active participation. Danish experts contribute to the United Nations' Sustainable Development Goals (SDGs) related to clean water and sanitation, and the country supports international research programs and conferences to share experiences and learn from global best practices.

The transformative potential of holistic management, advanced technologies, and public engagement is exemplified in Denmark's approach to water conservation. By prioritizing groundwater protection, optimizing water supply systems, and encouraging water-saving practices, Denmark has achieved notable success in reducing water consumption and promoting sustainable water use. Moreover, its active role in international collaboration and knowledge sharing has solidified Denmark's position as a global leader in sustainable water management, inspiring and assisting other countries address their water challenges and strive toward water security.

Access to More Working Knowledge for Better Communication

1. Read the scenario given below and discuss in pairs or in a group of three or four about how water diplomats can adjust actions in order to promote water saving.

Pakistan's water resources are heavily dependent on transboundary rivers, with over 70% of its surface freshwater originating from or transiting through neighboring countries such as India and Afghanistan. These rivers play a critical role in Pakistan's water security, agriculture, and economic development. Meanwhile, Pakistan has a financially unsustainable irrigation system because water pricing is not linked to the actual volume of water supplied to farmlands. This disconnect results in significant water wastage by farmers. Without incentives to conserve water, farmers tend to flood their entire fields using traditional flood irrigation methods rather than adopting high-efficiency irrigation systems (HEIS). Consequently, water productivity remains low.

2. Read the scenario given below and discuss what you will do as a potential water diplomat to promote the water saving campaign.

In 2008, the Israeli government launched a major water conservation campaign to counter water scarcity. The focus was on installing water-saving devices in bathrooms, toilets, and kitchens. Low-flow showerheads for bathrooms, dual-flush toilets, and faucet aerators for kitchens were promoted. These devices, like low-flow showerheads with aerated nozzles, use less water without sacrificing functionality. Dual-flush toilets let users choose flush volumes, thus reducing waste. Faucet aerators mix air with water, cutting flow.

The goal was to reach over 50% of all households, public buildings, and government offices. To achieve this, the government ran a public awareness program. It used media, public service announcements, and outreach in schools.

3. Read the information given below and discuss as a potential water diplomat where you can start to promote water conservation in an international setting by learning from the EU.

The EU's water diplomacy with non-EU member states encompasses several crucial elements:

The EU actively fosters cooperation regarding transboundary water bodies. It collaborates with non-EU states that share rivers, lakes, or aquifers to formulate joint management strategies. For instance, the Danube River Basin encompasses numerous non-EU countries. Here, the EU spurs coordinated endeavors to guarantee sustainable water utilization, effective pollution control, and robust flood protection. By bringing together different stakeholders, the EU aims to create a unified approach to manage these vital water resources, taking into account the diverse needs and interests of all parties involved.

The EU offers technical expertise and financial backing to empower non-EU states to enhance their water management capabilities. This support can manifest in various forms, such as assisting in the development of water infrastructure. This might involve building new water treatment plants, improving water distribution networks, or constructing dams for water storage. The EU also helps in implementing advanced water treatment technologies, enabling non-EU states to purify water more efficiently and meet higher quality standards. Additionally, it contributes to enhancing water resource monitoring systems, providing the tools and knowledge necessary to accurately assess water quantity and quality, and make informed decisions about water management.

The EU engages in in-depth policy dialogue with non-EU states to encourage the adoption of comparable water policies and regulations. The overarching goal is to establish consistent standards across borders for water quality, conservation, and pricing. For water quality, this means promoting the implementation of strict pollution control measures and water quality monitoring protocols. In terms of conservation, the EU advocates for policies that encourage the efficient use of water resources and the protection of water-related ecosystems. Regarding pricing, the EU aims to help non-EU states develop rational water pricing mechanisms that reflect the true value of water and encourage sustainable consumption. By achieving such policy harmonization, cross-border cooperation in water management becomes more seamless and effective.

The EU may present funding opportunities or attract investment for water-related projects in non-EU states. This can involve joint initiatives focusing on improving water supply and sanitation. For example, financing the construction of wells, pipelines, and sanitation facilities in areas with limited access to clean water and proper sanitation. It can also include projects aimed at enhancing water efficiency in agriculture and industry. In agriculture, this could mean promoting the use of drip irrigation systems or other water-saving agricultural techniques. In industry, it might involve supporting the implementation of water-recycling and reuse technologies. By facilitating these funding and investment opportunities, the EU helps non-EU states to develop and upgrade their water-related infrastructure and practices.

Further Recommended Reading

[1] Tenne, A. *Sea Water Desalination in Israel: planning, coping with difficulties, and economic aspects of long-term risks*. Israel Water Authority, State of Israel Desalination Division, 2010.

[2] 胡文俊,孙岩. 世界主要国家的节水措施及其启发与思考. 水利经济, 2024,42(2): 1-6.

[3] Marin, P., Tal, S., Yeres, J., et al. *Water management in Israel: key innovations and lessons learned for water-scarce countries*. Washington D. C.: World Bank, 2017. https://documents.worldbank.org/en/publication/documents-reports/documentdetail/657531504204943236/water-management-in-israel-key-innovations-and-lessons-learned-for-water-scarce-countries.

[4] PUB Singapore's National Water Agency. *Our water story*. 2024. https://www.pub.gov.sg/Public/WaterLoop/OurWaterStory.

Unit 13 Agricultural Water Engineering

Background Study for Successful Intercultural Communication

Read authentic Chinese and English topical texts in advance before making a glossary for their associated terminology and jargon as follows.

irrigation system 灌溉系统
drainage system 排水系统
soil moisture 土壤湿度
water conservation 节约用水
crop yield 农作物产量
surface irrigation 地面灌溉
drip irrigation 滴灌
flood irrigation 大水漫灌
groundwater recharge 地下水补给
water footprint 水足迹
salinity management 盐碱地管理
rainwater harvesting 雨水收集
aquifer 含水层
hydroponics 水培
sustainable agriculture 可持续性农业
water quality 水质
agricultural runoff 农业径流
climate resilience 气候变化适应能力
drought resistance 抗旱能力
water-use efficiency 水利用效率
agricultural engineering 农业工程
fertigation 施肥灌溉

water distribution network 水分配网络
irrigation scheduling 灌溉调度
erosion control 侵蚀控制
watershed management 流域管理
precision agriculture 精准农业
soil erosion 土壤侵蚀
water storage 水储存
irrigation efficiency 灌溉效率
agricultural practices 农业实践
bioengineering 生物工程
micro-irrigation 微灌溉
crop rotation 作物轮作
water resource assessment 水资源评估
evapotranspiration 蒸散
wetlands management 湿地管理
agricultural technology 农业技术
remote sensing 遥感技术
soil conservation 土壤保护
agricultural policy 农业政策
nutrient management 营养管理
climate-smart agriculture 气候智能型农业
water pricing 水资源定价
agricultural sustainability 农业可持续性
sedimentation 沉淀
crop irrigation 农田灌溉
water-efficient crops 节水作物
aquaponics 水产养殖结合种植
field mapping 地块测绘
land reclamation 土地复垦
water-sensitive urban design 水敏感城市设计
agricultural biodiversity 农业生物多样性

Text A
California's Approach to Agricultural Water Management

Pre-reading Questions

1. What are the main agricultural water engineering projects in California, and how have they been designed to meet the water needs of different cultural and economic groups?

2. Some people might use comparisons to show that even in a state with advanced water management systems, excessive agricultural water use can lead to problems such as groundwater depletion and environmental degradation. Do you agree with them or not?

3. How do cultural attitudes towards agriculture and infrastructure in California impact the implementation and acceptance of water engineering solutions?

4. What are the major communication barriers between water engineers and farmers in California, considering their different cultural backgrounds and professional languages? And how can agricultural water engineering projects in California better communicate with local communities to address concerns about environmental impacts and cultural heritage?

5. How do water shortages and agricultural engineering solutions in California affect different cultural communities, and what are the unique challenges and opportunities for each group? In the process of water engineering project implementation, how can the concerns and needs of various cultural communities in California be better addressed through effective communication?

6. What role does intercultural communication play in promoting cooperation among different stakeholders in California's water engineering and agricultural development?

7. What can other regions or countries learn from California's experience in integrating culture and water management in agriculture?

California's agricultural sector plays a vital role in the state's economy, accounting for over one-third of the United States' vegetables and two-thirds of its fruits and nuts. This remarkable productivity, however, comes with significant water consumption, highlighting the crucial role of agricultural water engineering in achieving sustainable resource management.

One of the underlying reasons is that it has implemented a range of comprehensive policies aimed at enhancing water use efficiency in agriculture. The Sustainable Groundwater Management Act (SGMA) enacted in 2014 addresses the pressing issue of groundwater over-extraction by requiring local groundwater sustainability agencies to devise plans that ensure sustainable water use through balanced withdrawals and replenishment. Another key initiative, the California Water Action Plan, prioritizes water conservation, modernization of water infrastructure, and ecosystem restoration to create a more resilient water system. This strategic plan promotes advanced irrigation techniques and incentivizes water-saving innovations. Furthermore, California's Department of Water Resources supports farmers through grants and loans, funding programs like the Agricultural Water Use Efficiency (AWUE) Grant to enhance irrigation efficiency and boost water recycling efforts.

In addition, technological innovation has been a cornerstone of California's efforts to optimize water use in agriculture. Drip and micro-irrigation systems, which deliver water directly to plant root zones, significantly reduce evaporation and runoff. California is one of the first states in adopting drip irrigation, especially for high-value crops such as almonds and grapes. Precision irrigation systems equipped with soil moisture sensors enable real-time monitoring of soil conditions, ensuring efficient water application. Satellite and remote sensing technologies further enhance decision-making by providing critical insights into crop water needs and soil health, enabling tailored irrigation practices. In addition, California's investment in advanced water recycling plants has been transformative, allowing treated wastewater to be repurposed for agricultural irrigation and reducing dependency on limited freshwater resources.

California's expertise in agricultural water engineering has established it as a global leader in water management. A notable example is its partnership with Israel, a pioneer in drip irrigation technology, which has fostered the exchange of knowledge and the development of hybrid systems tailored to both Californian and

Israeli agricultural needs. Additionally, California's experts actively contribute to global conversations on sustainable water use through participation in organizations such as the World Water Council and international water summits. The state's collaborations with developing countries further underscore its commitment to sharing technological expertise and offering training programs on efficient irrigation and water conservation practices.

Despite significant advancements, California continues to face challenges, including the impacts of climate change, aging infrastructure, and issues of water equity. Addressing these challenges will require an expanded focus on groundwater recharge projects to counter overdraft and greater international cooperation to develop climate-resilient water management systems. Moreover, advancing research into drought-resistant crop varieties will be essential to ensuring sustainable agricultural practices in the face of future uncertainties.

California's approach to agricultural water engineering reflects a balanced integration of technology, policy, and international collaboration. By continually innovating and sharing its expertise, the state not only addresses its own water challenges but also contributes to global efforts to achieve sustainable water management.

Access to More Working Knowledge for Better Communication

1. Study the water-saving methods with their corresponding benefits. Then discuss the diplomatic implications and strategies for promoting these methods in international cooperation.

Water-Saving Methods	*Benefits*
Drip Irrigation	Minimizes evaporation by delivering water only to the roots of the plant.
Soil Moisture Sensors and Precision Irrigation	Uses real-time data to ensure that water is applied only when necessary.

Conservation Tillage	Conserves soil moisture by reducing land disruption.
Recycled Water	Reduces reliance on freshwater through the reuse of treated wastewater.
Water-Efficient Crop Varieties	Focuses on developing crops that need less water to grow.
Storage and Management	Improves infrastructure to ensure water storage during dry periods.
Climate-Smart Agriculture	Mitigates the effects of changing weather patterns on water resources and agriculture.

2. Explore additional information from international organizations regarding the urgent challenges facing the agricultural industry, and reflect on how you, as a future water diplomat, can contribute to addressing these issues.

—To sustain an estimated global population of 10 billion by 2050, agricultural output must rise by nearly 50% from 2012 levels (FAO, 2017), with a significant portion of this growth anticipated through improved irrigation, water capture, and storage technologies, along with other innovations. However, increasing water demands for domestic, industrial, and urban use, driven by wealth-related changes in diets and economic growth, also need to be addressed.

—Rising water demand, combined with the effects of climate change, presents a major challenge for all countries regarding water allocation and governance, especially in relation to food security. On one hand, it is crucial to balance secure access to water resources with the required infrastructure investments. On the other hand, institutions and infrastructure operations must remain flexible enough to allow water to be reallocated transparently and equitably in response to climate change effects and shifting needs. Effective governance of water tenure is key to achieving this balance.

—Effective strategies require a unified approach to land and water management within the agricultural sector, while also addressing concerns such as disaster risk reduction, biodiversity restoration, and the promotion of sustainable livelihoods for communities. It is essential to strengthen coordination and joint governance of water and climate to integrate freshwater issues into climate mitigation and adaptation planning, especially when setting nationally determined contributions (NDCs). This would ensure the active participation of the food and agriculture sector, along with relevant ministries and stakeholders, in developing cohesive solutions. Governance frameworks and tools must be tailored to local contexts. Enhanced coordination and cooperation among stakeholders, sectors, and transboundary basins will be crucial in managing trade-offs effectively (United Nations, 2022).

—Innovations in digital agriculture can enhance farmers' productivity, profitability, and affordability by improving integrated information systems tailored to different crops and agroecological zones. These systems help manage water, soil, and land resources while also strengthening governance, institutions, and human capital by providing tools for better decision-making. Existing global platforms, such as the FAO Water Productivity Open-access portal (WaPOR), Global Agro-ecological Zones (GAEZ) geospatial mapping, the Global Information and Early Warning System on Food and Agriculture (GIEWS), and the Agricultural Stress Index System (ASIS) for drought, soil, and irrigation mapping, along with land suitability tools, can support the efficient and sustainable management of natural resources.

Text B
Cooperation with Israel in Dripping Irrigation Technology

Pre-reading Questions

1. In some cultures, traditional irrigation might be deeply rooted in history and tradition, seen as a way of life and connection to the land. Understanding this cultural significance is crucial for water professionals when communicating with stakeholders who might be resistant to change. How can future water diplomats use this knowledge to approach the introduction of drip irrigation in a way that respects

and builds on the existing cultural values?

2. Clear and persuasive communication across cultures is crucial for understanding the role of international collaboration in spreading drip irrigation technology. With experiences in international seminars or workshops, water diplomats can better present the technology to people from different countries and cultural backgrounds by adapting their communication style, using simple and understandable language, and providing visual aids if necessary to ensure that the message is effectively conveyed. Do you agree with that?

3. Some water diplomatic challenges might be cultural in nature. There could be resistance from some farmers who are used to the old ways of doing things. Do you think it is possible to approach such challenges with cultural sensitivity, by involving the local community in the problem-solving process, listening to their concerns, and co-creating solutions that take into account their cultural and practical needs?

4. How does Israel's drip irrigation technology contribute to environmental protection and sustainable agriculture in different countries? What are the challenges and solutions in promoting sustainable water use through this technology?

5. What training programs does Israel offer to help other countries' technicians and farmers master drip irrigation technology? How effective are these programs in facilitating knowledge transfer across cultures?

6. How does Israel customize its drip irrigation technology for different international markets in terms of climate, soil types, and crop varieties?

Situated in an arid region with limited freshwater resources, Israel has faced the persistent challenge of optimizing water use in agriculture. In response, the country developed and implemented drip irrigation technology—a method that revolutionizes water delivery to crops by supplying precise amounts directly to plant root zones. This breakthrough has transformed water management in agriculture, making it possible to grow high-yield crops even in water-scarce regions.

Drip irrigation was invented in the 1960s by Simcha Blass and his son Yeshayahu Blass. Unlike conventional irrigation methods such as sprinklers or flood irrigation, drip irrigation delivers water slowly and directly to the root zones through a network of pipes, tubes, and emitters. This minimizes water loss through evaporation and runoff, enhances nutrient absorption, and reduces weed growth by targeting water application only where needed.

Modern drip irrigation systems are now equipped with advanced technologies, including pressure-compensating drippers, which ensure a consistent water flow regardless of elevation or pressure changes. Integrated fertigation systems deliver nutrients along with water, enhancing crop growth and reducing fertilizer wastage. IoT and automation technologies use smart irrigation controllers and real-time data from sensors to optimize water delivery schedules based on soil moisture, weather forecasts, and crop requirements.

Drip irrigation has found widespread use across diverse agricultural sectors in Israel. High-value crops such as citrus, pomegranates, and olives benefit significantly from precise water and nutrient delivery. Vegetables like tomatoes, peppers, and cucumbers achieve higher yields and quality through drip systems. Controlled environments like greenhouses, paired with drip irrigation, reduce water consumption and maximize productivity. Even traditional crops like wheat and cotton can be grown efficiently under drip irrigation. The technology's efficiency has enabled farmers to increase yields by up to 50% while using 30% ~ 50% less water than conventional irrigation methods.

Israel's government has played a pivotal role in the adoption and advancement of drip irrigation through policies and initiatives. The government provides financial incentives to farmers for adopting water-efficient technologies like drip irrigation. Strict water management laws prioritize water conservation, including mandating the use of efficient irrigation systems in agriculture. Public and private sectors collaborate through organizations such as the Volcani Center to advance irrigation technology and practices. Progressive pricing policies charge farmers for water based on usage levels, incentivizing efficient practices.

Israel's expertise in drip irrigation technology has already been shared with countries worldwide through training programs, partnerships, and international aid. Companies like Netafim, founded in Israel, are global leaders in manufacturing and deploying drip irrigation systems. It has been at the forefront of providing innovative solutions for water-efficient agriculture across Africa, the Middle East, and other water-scarce regions. In Africa, it has projects in countries like Kenya, Ethiopia, and Morocco. In the Middle East, it has been active in Jordan, Palestine, and several Gulf countries. Netafim offers a wide range of drip irrigation products, from simple drip emitters to advanced automated irrigation systems. These systems are designed to be adaptable to different soil types, crop varieties, and water sources. The company also provides comprehensive training services, teaching local farmers and technicians how to install, maintain, and optimize these systems for maximum water conservation and agricultural productivity.

Rivulis, another major Israeli company in the water-efficient irrigation sector, has expanded its operations globally, targeting regions with water scarcity. In the Middle East, it has contributed to the development of sustainable agricultural practices in countries such as Iraq and Oman. In Africa, it has been involved in projects in Egypt and Sudan. Rivulis' portfolio includes driplines, sprinklers, and filtration systems. The company also provides technical support and training, enabling local communities to make the most of their water resources through proper irrigation management.

Programs such as Mashav, Israel's Agency for International Development Cooperation, provide training to farmers and policymakers in water management. Mashav's training programs are comprehensive and far-reaching. For farmers, it offers hands-on training sessions on the latest water-saving techniques. These include in-depth instruction on the installation, maintenance, and operation of drip irrigation systems. Farmers are taught how to customize these systems according to the specific needs of their crops, soil types, and water availability. For example, in arid regions, they learn how to adjust the flow rate and frequency of water delivery to minimize evaporation and ensure that plants receive the optimal amount of water.

Policymakers also benefit greatly from Mashav's programs. They are provided with insights into effective water-management policies, such as water-rights allocation, water-pricing mechanisms, and regulatory frameworks for sustainable water use. Through case studies and interactive workshops, policymakers around the world can understand how Israel has managed its limited water resources to support a thriving agricultural sector. This knowledge equips them to make informed decisions and design policies that promote water conservation and agricultural development in their own countries.

Collaborative projects with countries such as India, Kenya, and China have introduced drip irrigation in water-scarce areas, boosting their agricultural productivity and food security. In India, where water scarcity is a major challenge for agriculture, the introduction of Israeli drip irrigation technology has been transformative. These projects have been implemented in states like Rajasthan and Gujarat. By using drip irrigation, farmers in these areas have been able to increase their crop yields significantly. Crops such as cotton, which is water-intensive, now can be grown more efficiently with less water wastage. This not only boosts agricultural productivity but also contributes to food security at the local and national levels.

In Kenya, similar collaborative efforts have been underway. Israel's expertise has been instrumental in helping Kenyan farmers in arid and semi-arid regions. These areas, which are prone to drought, have seen a shift towards more sustainable agricultural practices. Drip irrigation systems have been introduced in small-scale farming communities, enabling them to grow a wider variety of crops throughout the year. This has improved the livelihoods of farmers and reduced their dependence on erratic rainfall patterns.

China, with its vast agricultural landscape and diverse climate zones, has also engaged in productive cooperation with Israel. Joint projects have focused on developing and adapting irrigation technologies to different regions in China. In the northwestern regions, where water is scarce, drip irrigation and other water-saving techniques have been introduced. These projects have not only increased crop yields but also contributed to the overall sustainability of China's agricultural sector.

Despite its success, drip irrigation faces challenges such as high initial costs, the need for skilled operation, and maintenance demands. To overcome these barriers, Israel continues to innovate, focusing on developing affordable solutions for smallholder farmers, advancing solar-powered irrigation systems to reduce energy costs, and enhancing precision agriculture tools that integrate AI and machine learning for better water management.

Access to More Working Knowledge for Better Communication

1. Read the information from UN and FAO reports listed below and discuss in pairs or at a mock negotiating session about how water resources is used and allocated in the agricultural industry.

—Globally, 84% farms are small (less than 2 ha), they operate around 12% of all agricultural land but produce roughly 35% of the world's food by value (Lowder et al., 2021). Women are often responsible for balancing household water use to agricultural water needs, and securing water typically falls on women and girls, adding to their unpaid and domestic work burdens. There needs to be "a fundamental shift beyond considering water as a resource for food production to focusing on people and the role that water plays in their livelihood strategies" (WWAP, 2006).

—Improvements in irrigation, combined with water reuse, desalination and groundwater recharge, are important strategies for transforming agricultural water management. Additionally, the digitalization of the agricultural sector is increasing the availability of accurate and real-time data for farmers, allowing them to make better decisions on crop choices and water use. Innovations in water-scarce regions have allowed farmers to better control water and fertilizer inputs to improve crop yields. Traditional drip irrigation systems, requiring higher capital and energy cost, may not be suited for smallholder farmers in the NENA region with plot sizes ranging from 0.09 ha to 0.76 ha. Innovative projects have introduced valves that provide drip irrigation at very low pressures of 0.15 Bar versus the 0.50 to 1.00 Bar needed for a conventional system. These installations use a smaller-capacity pump and a rooftop water tank, allowing for energy savings of up to 50%.

2. Identify people or institutions mentioned in the following passage and talk about what they have done in intercultural communication with interested parties on global water issues.

Nearly 84% of smallholder farms in low-and-middle-income countries are situated in water-scarce regions, and fewer than one-third have access to irrigation. There is a pressing need to focus on providing support to smallholder farmers, particularly women and children, as their role is crucial for achieving the Sustainable Development Goals (SDGs) and conserving local ecosystems. The FAO's framework on extreme rural poverty emphasizes that conserving and restoring natural resources should directly benefit the rural poor, especially those in remote and marginalized areas. This connects to promoting responsible governance of resource tenure. Acknowledging the legitimate tenure rights of people to use, manage, and control land, water, biodiversity, forests, and fisheries is vital for helping the rural extreme poor adapt to climate change.

Increasing awareness of the role of water in rural livelihoods and adopting participatory approaches centered around smallholder farmers can help focus actions on building resilience, identifying and adapting water technologies, and promoting smart investments in water for poverty reduction. International and national partners are developing methodologies such as livelihood mapping to aid investors and policymakers in prioritizing, planning, and implementing water-related interventions to support smallholder farmers. These efforts include conducting regional and national studies on rural poverty reduction through water-related actions, as well as applying water technologies to enhance the impact of development projects focused on poor farmers, with a particular emphasis on women.

Further Recommended Reading

[1] United Nations. *The United Nations World Water Development Report 2024: water for prosperity and peace*. Paris: UNESCO, 2024. https://unesdoc.unesco.org/ark:/48223/pf0000388948.

[2] Cooley H, Gleick P, and Wilkinson R. *Agricultural water conservation and efficiency potential in California-issue brief*. National Resources Defense Council, 2014. https://pacinst.org/publication/agricultural-water-conservation-and-efficiency-potential-in-california-issue-brief/.

[3] California Natural Resources Agency, California Environmental Protection Agency, California Department of Food and Agriculture. *California water action plan*. 2014. https://resources.ca.gov/CNRALegacyFiles/docs/california_water_action_plan/2014_California_Water_Action_Plan.pdf.

[4] Aivazidou, E. *Water management in agriculture and industry: challenges, trends, and opportunities*. Sustainability MDPI, 2022, 14:6.

[5] National Water Reuse Action Plan. *From water stressed to water secure: lessons from Israel's water reuse approach*. 2022 US Delegation Summary, 2023. https://www.epa.gov/system/files/documents/2023-03/From%20Water%20Stressed%20to%20Water%20Secure%20-%20Lessons%20from%20Israel%27s%20Water%20Reuse%20Approach.pdf.

Unit 14 Management of Hydraulic and Hydropower Projects

Background Study for Successful Intercultural Communication

Read authentic Chinese and English topical texts in advance before making a glossary for their associated terminology and jargon as follows.

grid integration 电网整合
hydraulic structures 水工结构
community involvement 社区参与
energy storage solutions 能源储存解决方案
emergency response planning 应急响应规划
water management policies 水管理政策
hydraulic engineering 水利工程
hydropower generation 水力发电
project feasibility study 项目可行性研究
environmental impact assessment 环境影响评估
dam construction 大坝建设
reservoir management 水库管理
flood control 防洪
renewable energy 可再生能源
cost-benefit analysis 成本效益分析
risk management 风险管理
energy efficiency 能源效率
hydrological modeling 水文模型
climate change adaptation 气候变化适应
sediment management 淤泥管理

aquatic ecosystems 水生生态系统
power transmission 电力传输
policy framework 政策框架
maintenance strategies 维护策略
monitoring and evaluation 监测与评估
licensing and permits 许可和许可证
funding sources 资金来源,资助来源
infrastructure development 基础设施建设
technological innovation 技术创新

Text A
Global Efforts Behind the Mount Coffee Hydropower Plant

Pre-reading Questions

1. Cultural and economic factors, such as local attitudes towards renewable energy and the cost of infrastructure development, might influence a country's diplomatic decisions for hydropower. What do you think water diplomats can do to help African nations to understand their specific energy needs and understand how hydropower can fit into their long-term energy plans?

2. How do you think an experienced water diplomat would adjust his or her intercultural communication strategies to navigate the complex negotiations between states, local governments, and interest groups when transitioning from working on water-management projects in a highly centralized country with efficient top-down policies to a federal country like the United States?

3. When a qualified water diplomat talks with local community members, they will approach discussions with respect for local cultures and values. Instead of a top-down approach, they would engage in a more consultative and inclusive manner, perhaps by involving local elders or community leaders in the initial stages of planning and information-sharing. This could mean holding community meetings in a traditional setting, like a village square, to make the locals feel more comfortable and involved. What's your opinion on that?

4. In the context of the Mount Coffee project, there are complex technical terms related to hydropower. Water diplomats' work requires not only language skills but also the ability to simplify complex ideas. How could future water diplomats prepare themselves for making the information more relevant to stakeholders?

5. It is important to guarantee that all stakeholders' voices are heard and their interests are considered when dealing with international organizations, government agencies, and non-governmental organizations from various cultural backgrounds in a water-cooperation initiative. Do you think the statement is applicable for the rebuilding of Mount Coffee Hydropower Plant in Liberia?

6. When advocating for the adoption of international water-management standards in a region with a different cultural approach to environmental and water governance, what communication strategies would you use to gain support?

The Mount Coffee Hydropower Plant in Liberia is a perfect illustration of international collaboration for sustainable energy development and management in a resource-scarce context. Located along the Saint Paul River in western Liberia, this hydroelectric facility is central to Liberia's efforts to address its energy challenges. After years of civil unrest in the 1990s, which resulted in the widespread destruction of the Mount Coffee Hydropower Facility, the plant has undergone a comprehensive rebuilding process. The facility, once a key source of electricity for Liberia, was severely damaged during the country's prolonged period of conflict. This left the nation with a substantial gap in its energy infrastructure, further complicating its efforts to support economic growth and development.

Key partners, including Statkraft from Norway, the U. S. Agency for International Development (USAID), China, and the African Development Bank, provided vital financial, technical, and construction support to restore, manage and operate the hydropower plant.

At the heart of the Mount Coffee Hydropower Plant's management is the Liberian government. The Liberian government provides oversight and regulatory support, ensuring that the plant operates in compliance with the nation's energy policies and priorities. As part of Liberia's development plan, the government views the plant as an essential piece of infrastructure for achieving energy security, facilitating economic growth, and improving access to electricity for its population. It is also committed to ensuring that the benefits of the plant reach local communities through expanded access to clean energy, particularly in rural areas.

Norway has been a major partner in the rehabilitation and modernization of the Mount Coffee plant, primarily through its leading renewable energy company, Statkraft. Norway's involvement began when the country's government, under its foreign aid and development strategy, identified Liberia as a priority for sustainable energy development. Statkraft's role was integral during the rehabilitation phase, providing critical expertise in hydropower operations, from initial assessments to technical support in modernizing the plant's systems and optimizing energy production. As one of the largest renewable energy companies in Europe, Statkraft's vast experience with hydropower has helped Liberia maximize the plant's energy output while minimizing environmental impacts.

U.S. Agency for International Development (USAID) has played a vital role in supporting the Mount Coffee Hydropower Plant's management. The U.S. government agency's involvement was crucial, especially after the devastating effects of the Ebola crisis, which put further pressure on Liberia's energy needs. USAID's assistance was key to restarting and upgrading critical energy infrastructure. The agency provided financial support as well as technical assistance in operational strategies, including the design and commissioning of various components of the hydropower system. Through its long-term partnership with the government of Liberia, USAID aims to improve access to electricity, stimulate economic growth, and enhance the sustainability of Liberia's energy sector.

China has also been a significant player in the infrastructural development of Liberia, including energy projects like the Mount Coffee Hydropower Plant. Through the expertise of Chinese firms, construction work related to the rehabilitation of the hydropower plant and installation of necessary infrastructure was made possible. Chinese companies have provided critical materials, machinery, and technology for plant construction and upgrade, supplementing the international efforts aimed at reactivating the plant. Additionally, Chinese financial institutions have offered loans and investment that have facilitated the overall development and expansion of the facility.

The rehabilitation and expansion of the Mount Coffee Hydropower Plant would not have been possible without financial support from international financial institutions, such as the African Development Bank (AfDB) and the World Bank. These institutions provided loans and grants that were crucial for the plant's modernization. The AfDB, in particular, has focused on financing energy infrastructure projects across Africa, including Liberia, and worked with the government and other stakeholders to ensure that the Mount Coffee facility would be sustainable and affordable, and deliver reliable power to the national grid. Both the AfDB and the World Bank's engagement demonstrates how such global financial entities are essential partners in achieving energy transformation and sustainable development.

The success of the Mount Coffee Hydropower Plant also depends on local engagement. With assistance from international organizations, particularly Norwegian development agencies and the Liberian government, programs have been implemented to enhance local communities' understanding of energy production and environmental stewardship. Furthermore, workforce development programs and capacity-building efforts ensure that Liberians can operate, maintain, and benefit from the plant for years to come, facilitating knowledge transfer and self-sufficiency.

The Mount Coffee Hydropower Plant's management is a prime example of how international cooperation can drive development in sectors that are key to long-term sustainability. With contributions from Norway, the United States, China, the World Bank, the African Development Bank, and the Liberian government, the plant has gone through a complex rehabilitation process, aiming to provide more affordable and reliable electricity to the people of Liberia. Together, these diverse stakeholders are not only rebuilding the plant but also contributing to the country's energy resilience, providing Liberia with cleaner and sustainable power to support both its current needs and future growth.

Access to More Working Knowledge for Better Communication

1. Read about some of the challenges in the operation and management of the Mount Coffee Hydropower Plant provided by the World Bank. Then, discuss in pairs or at a mock negotiating session about how water diplomats can make a difference by bringing the necessary resources of different countries together.

Challenge 1: Human resources and capacity building

The most challenging issue regarding capacity building for the operation and management (O&M) of the Mount Coffee Hydropower Plant, as in many developing countries, is building and retaining human resources. Long-term sustainable O&M can only be achieved if the O&M is locally staffed, with managers and technicians who are duly trained and motivated.

The challenges with training staff at Mount Coffee include:

Affording the O&M contractor adequate and sufficient control and authority over the trainees through a formal agreement with the employer, the O&M contractor, and the seconded trainee.

Setting up an individual performance reward system to acknowledge the progress, commitment, and dedication of the trainees. It is recognized that this may conflict with the employment status and salary structures of local government agencies.

Retaining the newly trained staff, who may be attracted by other opportunities in different sectors.

The purpose of capacity building is to develop expertise and knowledge within the country, not just within the confines of the utility. Based on the experience gained from working at Mount Coffee, Hydro Operation International (HOI) proposed that the optimal solution could be to create a specific private local O&M company. This company would be staffed mainly with local personnel and managed by the O&M contractor for the purpose of operating and maintaining the plant in a sustainable way.

Challenge 2: Financing the operation and management

Multiyear budgets should be prepared in advance, with reserves allocated for major maintenance works and overhauls. The Mount Coffee plant is owned by Liberia Electricity Corporation (LEC) and will be managed by LEC after the end of the O&M contract. However, because the plant does not have its own budget, accounts, and control, the plant management lacks the freedom to operate the facility optimally.

The financial plan should allocate capital in a reserve fund to finance overhauls when they become necessary. In the absence of reserve funds, overhauls are typically not completed or are postponed, threatening the life of the equipment.

Some elements required for the O&M program at Mount Coffee, which are critical to ensuring safe and sustainable O&M, were not included in the project rehabilitation budget. These elements remain the responsibility of LEC to finance. Examples include a village, some major maintenance equipment (e. g., a truck with a jib crane, a mobile crane, or a fire truck), site communications, and modern IT facilities.

Creating a specific profit center for the plant, with its own management, financial and operational targets, and sources of revenue, may enable more optimal asset management. This could further guarantee the performance, reliability, and sustainability of the plant.

2. Identify existing international cooperation efforts in hydraulic and hydropower project management and carefully investigate the aspects of water diplomacy that future water experts should focus on.

—*Better Hydro: Compendium of Case Studies 2017* is a collaborative report between the International Hydropower Association (IHA) and the World Bank Group. This publication presents 34 case studies that highlight good practices in sustainable hydropower development. The compendium was officially launched at the 2017 World Hydropower Congress in Addis Ababa, Ethiopia. It was welcomed by stakeholders as an important resource for highlighting good practices in sustainable hydropower development, especially in Africa where hydropower is prioritized as a source of renewable energy. Richard Taylor, Chief Executive of IHA, emphasized that the Hydropower Sustainability Assessment Protocol enables project developers to identify gaps in their practices and processes, and understand how to address them. This compendium is a significant step towards sharing these examples widely, promoting better practices in the hydropower sector.

—The International Hydropower Development Conference (IHDC) 2024 took place in Beijing, China, from October 29 to 31, 2024. The conference focused on hydropower and renewable energies, with 48 papers presented and a significant emphasis on international cooperation and sustainable development.

—Climate-proof Watershed Management Design and Resilience is a joint project by the Food and Agriculture Organization (FAO) and the United Nations Economic and Social Commission for Western Asia (ESCWA) focuses on developing climate-proof watershed management plans and resilience packages. This project aims to improve water resources management in the Arab region, which is among the most water-scarce regions globally. Watershed management plans and resilience packages were developed for three watersheds in Lebanon and Algeria.

—The IEA Technology Collaboration Programme on Hydropower is a working group of International Energy Agency member countries and others interested in advancing hydropower worldwide. The program aims to encourage the sustainable use of water resources for hydropower development and management through awareness, knowledge sharing, and support.

—The International Network on Small Hydropower (INSH) promotes the development of small hydropower projects through international cooperation. The network focuses on standardization, training, and capacity building, particularly in developing countries. This initiative helps in sharing best practices and advancing sustainable hydropower development globally.

—The IHA plays a crucial role in promoting sustainable hydropower development through various initiatives, including the release of the 2024 World Hydropower Outlook. This annual publication tracks the progress of hydropower development globally and provides insights into policy, financial challenges, and good practices.

—The International Finance Corporation (IFC) has developed guidelines for environmental, health, and safety approaches in hydropower projects. These guidelines emphasize the importance of assessing potential environmental risks and impacts early in the project life cycle, designing projects with operational flexibility, and collaborating with stakeholders to develop watershed management programs.

Text B
The Grand Inga Project: DRC's Pivotal Role, International Partners, and Management Struggles

Pre-reading Questions

1. What specific communication strategies can be used to show respect for the other party's culture and thus build a good foundation for cooperation when facing cultural barriers in promoting international hydropower project collaboration?

2. How do archaeologists, historians, and local cultural groups communicate the significance of the cultural heritage sites near the dam to dam managers when dam management activities such as water-level fluctuations threaten the sites? What cultural-specific arguments can be used to emphasize the importance of preservation?

3. Intercultural competence is crucial for fostering sustainable water management practices as it helps water diplomats to build trust, negotiate effectively, and implement solutions. How do future water diplomats find common ground and explore win-win solutions to make the collaboration work by making sure the interests of both parties?

4. Negotiation processes are subject to various interpretations. Diplomats therefore need to ensure that they possess an open mind so that they do not impose their own cultural rigidity or preferences on others. Do you agree with the statement?

5. How do international companies involved in management of hydraulic and hydropower projects in foreign countries communicate with local communities and stakeholders? How can they address the concerns of the local people regarding potential future impacts on their lives and the environment?

6. What are the effective communication channels and mediation methods to resolve misunderstandings or conflicts between the project management and the

local community caused by cultural differences in the perception of project benefits and responsibilities?

The Grand Inga Hydroelectric Project on the Congo River stands as a beacon of hope for revolutionizing Africa's energy landscape. The Congo River, renowned for its prodigious water flow, ranks among the world's most powerful waterways in terms of hydro-power generation potential. This natural resource represents an untapped wellspring of energy, and the project's overarching goal is to harness it to produce an astronomical amount of electricity, far surpassing the capacity of many existing hydroelectric plants globally.

Envisioned as a multi-phase development even before the construction commenced, the Inga Falls on the Congo River lie at the heart of this ambitious initiative. Once fully developed, the project has the theoretical capacity to generate over 40,000 MW of electricity. This colossal output could serve as a lifeline for the energy-starved Democratic Republic of the Congo (DRC) and also provide power to neighboring African countries. Such a substantial power supply has the potential to fuel industries, uplift living standards, and propel economic growth across the entire African continent.

As the custodian of the land and water resources, the DRC government plays a pivotal role. It provides the physical expanse where the dams, power plants, and all associated infrastructure will be erected. Given that the Congo River's water is the cornerstone of the project, the DRC's authorization is non-negotiable for any development along its banks.

The government also formulates a comprehensive set of policies. These include land-use policies that determine how the project-related infrastructure will be sited, environmental regulations to mitigate potential ecological damage, and laws governing energy production and distribution. By establishing these rules, the DRC ensures that the project aligns with both national development aspirations and international best practices. To demonstrate its commitment, the DRC government planned to invest around \$4 billion to kick-start the construction process, which is a significant step towards the project's realization.

In 2009, the World Bank made a significant contribution by pledging a \$50 million grant. This funding was earmarked for environmental impact assessment (EIA) studies, which are crucial for understanding and minimizing the project's ecological footprint. It also supported technical feasibility studies to ensure the project's technical viability, officer training to build local capacity, the establishment of the Inga III Dam Authority to manage the project's operations, and the development of a communication plan to keep stakeholders informed.

The African Development Bank (AfDB) has been an active partner since 2010, providing \$15 million for feasibility studies of the Grand Inga and Inga III hydro projects. In 2013, it further solidified its support by signing two grant agreements with the DRC government, totaling \$5.25 million, for technical assistance in the development of Inga. These contributions are instrumental in building the technical and institutional capabilities necessary for the project's success.

The European Investment Bank, as a potential financier, holds the promise of injecting large-scale capital. Its involvement could be a game-changer, helping to bridge the substantial funding gap required for the project's construction, day-to-day operation, and long-term maintenance.

Ivanhoe Mines Energy DRC, in collaboration with the DRC's state-owned power company SNEL, has extended financing agreements for the upgrade of turbines at the Inga II hydroelectric complex. This showcases the private sector's role in the continuous operation and improvement of existing infrastructure related to the overall project.

International consortia from various countries, such as the one formed by the China Yangtze Three Gorges Group and Sinohydro Corporation, along with those from Republic of Korea, Canada, and Spain, have participated in the bidding process for the Inga III project. These consortia bring a wealth of advanced construction technologies, sophisticated project management expertise, and the necessary capital to execute large-scale infrastructure projects. Their participation is crucial for ensuring high-quality construction and efficient project implementation.

Concerns have been raised regarding the lack of transparency in the project's decision-making processes. Information about financial transactions, the bidding procedures that determine which companies get involved in construction, and the distribution of benefits among stakeholders have not always been readily accessible to the public. This opacity can breed public mistrust and create an environment conducive to potential corruption risks.

The project has endured political interference from various levels within the DRC. Frequent policy changes, power struggles, and political instability pose significant threats. For instance, a change in leadership can lead to a shift in project priorities. This not only causes delays in the construction timeline but also inflates costs due to re-planning and re-negotiation processes.

Despite the financial contributions from the DRC government and international financial institutions, the project still grapples with substantial financial shortfalls. The estimated cost of the entire Grand Inga project is staggering, and raising the requisite funds remains a formidable challenge. The DRC's weak economic situation, compounded by high levels of debt, limits its ability to contribute additional funds. Moreover, attracting consistent private investment is difficult due to the project's inherent risks, such as political instability and technical uncertainties.

Although environmental impact assessments have been carried out, the actual implementation of mitigation measures often falls short. The project has the potential to cause significant environmental damage, including alterations to the river's natural flow, methane emissions from flooded areas due to the decomposition of organic matter, and the destruction of vital wildlife habitats. Socially, the resettlement of local communities has been fraught with difficulties. Conflicts have emerged over land rights and compensation, highlighting the need for more effective social management strategies.

Access to More Working Knowledge for Better Communication

1. Read the following case cited in *Better Hydro: Compendium of Case Studies 2017* and discuss in pairs or at a mock conference session about how we can mobilize actions in order to promote asset reliability and efficiency in international cooperation in energy projects.

The Nam Lik 1-2 Hydropower Project, located on the Nam Lik River in Vientiane Province, Laos, is a notable energy initiative. With an installed capacity of 100 MW, it features two 50 MW Francis turbines and 58 MVA generators from Hangzhou Resource Power Equipment Company, having a 25-year predicted lifespan.

Fully owned by the Nam Lik Power Company (NLPC), a joint venture of China International Water and Electric Corporation (CWE, 90% share) and Électricité du Laos (EDL, 10% share), CWE being a subsidiary of China Three Gorges Corporation, the project operates under a 25-year build-own-operate-transfer contract, licensed in 2007 and commissioned in July 2010. Its main structures include a rockfill dam, spillway, tunnels, powerhouse, switchyard, and a saddle dam.

Since commissioning, it has exceeded power purchase and generation targets, thanks to strict equipment performance monitoring and an efficient maintenance routine. NLPC uses automated and manual methods for asset monitoring, with engineers checking equipment every two hours and monthly data analysis for risk identification.

Asset management involves categorizing equipment for different replacement frequencies. Mechanical assets are replaced based on performance, and upgrades are planned as new technology emerges. Engineers measure the exact level of wear and tear to equipment in their routine monitoring, and determine the optimal time for replacement. NLPC replaces most mechanical assets based on the performance of the equipment rather than its predicted lifespan. The company also plans to implement efficiency improvements as new technology develops, rather than through a planned long-term programme of upgrades.

Environmentally, it is a qualified project with significant CO_2 emission reductions. Socially, it has contributed to local development, winning awards from both China and Laos. Overall, it serves as a model for effective asset management and sustainability in hydropower, emphasizing the value of cross-cultural cooperation between Chinese and Lao stakeholders.

2. Read another case illustrated in *Better Hydro: Compendium of Case Studies 2017* and discuss in pairs or at a mock conference session about how comprehensive monitoring of hydrological resources and analysis based on climate trends and climate change scenarios can provide reliable assessment of medium-and long-term water availability.

The Jostedal Hydropower Project in Norway showcases the significance of comprehensive hydrological resource monitoring and climate-based analysis. Located on the east of the Jostedalen valley, it harnesses runoff from a 144 km^2 catchment area at 1,200-meter elevation, with the Jostedøla River flowing through and into the Sognefjord.

Developed and operated by Statkraft, the project has a 288-MW capacity, generating 874 GWh annually, and serves the purposes of power generation and flood protection since its commissioning in 1989. Its key structures include the Styggevatn reservoir, storing one-and-a-half years' average inflow and fed by Kupvatn (storing three years' average inflow), and a power plant inside the mountain housing a single 288-MW Pelton turbine. There is also a second power station, Leirdøla, with a specific discharge route.

Hydrologically, Statkraft monitors river flow, reservoir levels, and snow/ice storage through gauges and sampling stations at different altitudes. Climate trends and change scenarios, based on 80 years of field measurements, are used to predict water availability. The project is designed to manage floods from glacier melt and rain, with reservoirs playing a crucial role in flood control.

In operational management, sophisticated models are used for generation planning, enabling adaptation to future changes. Runoff is efficiently managed to avoid spills and control floods, with model-predicted inflow guiding power plant operation. The project takes a long-term view, integrating climate change considerations to meet flood control and spill-avoidance goals.

Further Recommended Reading

[1] World Bank. *Operation and maintenance strategies for hydropower: six case studies*. Washington, D. C.: World Bank, 2020. https://documents1.worldbank.org/curated/en/573221581011972528/pdf/Six-Case-Studies.pdf.

[2] International Hydropower Association. *Better hydro: compendium of case studies 2017*. London: International Hydropower Association and World Bank Group, 2017.

[3] United Nations Economic and Social Commission for Western Asia (ESCWA), Arab Center for the Studies of Arid Zones and Dry Lands (ACSAD), and Food and Agriculture Organization of the United Nations

(FAO). *Climate-proof watershed management design and resilience synthesis report*. Beirut: ESCWA, 2022. https://www.unescwa.org/sites/default/files/pubs/pdf/climate-proof-watershed-management-design-resilience-english.pdf.

List of References

1. AL-MUQDADI S W, OMER M F, ABO R, et al. Dispute over water resource management—Iraq and Turkey[J]. Journal of Environmental Protection, 2016, 7(8): 1096-1103.
2. BISWAS B, GHUTE B B. Flood risk management: assessment and strategy [M]. Singapore: Springer, 2024.
3. ALI R N, FAIZ-UR-REHMAN, MAHMOOD-UR-REHMAN W. Indus Waters Treaty between Pakistan and India: from conciliation to confrontation[J/OL]. The Dialogue 10, 2015, 2 : 166-181. [2024-10-11]. https://qurtuba. edu. pk/thedialogue/The% 20Dialogue/10 _ 2/Dialogue _ April _ June2015 _ 166 - 181. pdf.
4. ALLAN J A. Water wars: conflict and cooperation along international borders [M]. New York: MIT Press, 2001.
5. BAIRD I G, SILVANO R A M, PARLEE B, et al. The downstream impacts of hydropower dams and indigenous and local knowledge: examples from thepeace-Athabasca, Mekong, and Amazon[J]. Environmental Management, 2021, 67: 682-696.
6. BALL P. The water kingdom: a secret history of China[M]. London: Bodley Head, 2016.
7. BARANYAI G. Institutions of transboundary water governance [M]// BARANYAI G. European water law and hydropolitics. Cham: Springer, 2019: 45-52.
8. BLUMSTEIN S, PETERSEN-PERLMAN J D. When the water runs dry: supporting adaptive governance in transboundary river basins [J]. Water International, 2021,46(3): 306-324.
9. CALIFORNIA DEPARTMENT OF WATER RESOURCES. Agricultural water use efficiency programs[R]. 2023.

10. CALIFORNIA NATURAL RESOURCES AGENCY, CALIFORNIA ENVIRONMENTAL PROTECTION AGENCY, CALIFORNIA DEPARTMENT OF FOOD AND AGRICULTURE. California water action plan [R/OL]. [2025 - 01 - 20]. https://resources. ca. gov/CNRALegacyFiles/docs/california_water_action_plan/2014_California_Water_Action_Plan. pdf.

11. CHOUDHURY E, ISLAM S, SUSSKIND L. Complexity of transboundary water conflicts: enabling conditions for negotiating contingent resolutions[M]. London: Anthem Press, 2018.

12. GABRIELYAN H. Water conflicts between Turkey, Syria and Iraq in 1980-2000[M]// GABRIELYAN H. Turkey's water policy as part of the political strategy. Wiesbaden: Springer VS,2024: 195-239.

13. GLEICK P H. Water in crisis: a guide to the world's freshwater resources [M]. Oxford: Oxford University Press, 2014.

14. GLEICK P H. The world's water 2008-2009[M]. London: Island Press, 2009.

15. GLOBAL HIGH-LEVEL PANEL ON WATER AND PEACE. "A matter of survival", the report of the Global High-Level Panel on Water and Peace draws a lot of attention [R/OL]. [2024-12-01]. https://www. genevawaterhub. org/news/matter-survival-report-global-high-level-panel-water-and-peace-draws-lot-attention.

16. SHUVAL H I. Water quality management under conditions of scarcity: Israel as a case study[M]. New York:Academic Press,1980.

17. INTERNATIONAL BOUNDARY AND WATER COMMISSION. Minute No. 323: extension of cooperative measures and adoption of a binational water scarcity contingency plan in the Colorado River Basin[R/OL]. [2025-01-12]. https://www. ibwc. gov/wp-content/uploads/2023/03/Min323. pdf.

18. INTERNATIONAL HYDROPOWER ASSOCIATION. Better hydro: compendium of case studies 2017[R]. London: International Hydropower Association and World Bank Group, 2017.

19. INTERNATIONAL HYDROPOWER ASSOCIATION. 2024 world hydropower outlook [R/OL]. [2025 - 01 - 12]. https://www. hydropower. org/

publications/2024-world-hydropower-outlook.
20. INTERNATIONAL HYDROPOWER ASSOCIATION. Hydropower sustainability assessment protocol[R]. London:International Hydropower Association, 2020.
21. IRVINE K, CHUA L, EIKASS H S. The four national taps of Singapore: a holistic approach to water resources management from drainage to drinking water[J]. Journal of Water Management and Modeling,2014,22(C375).
22. ISLAM S, SMITH K M. Interdisciplinary collaboration for water diplomacy: a principled and pragmatic approach[M]. New York: Routledge, 2020.
23. ISLAM S, SUSSKIND L E. Water diplomacy: a negotiated approach to managing complex water networks[M]. 1st ed. New York: Routledge, 2013.
24. NATIONAL WATER REUSE ACTION PLAN. From water stressed to water secure: lessons from Israel's water reuse approach[R/OL]. [2025-01-10]. https://www.epa.gov/system/files/documents/2023-03/From%20Water%20Stressed%20to%20Water%20Secure%20-%20Lessons%20from%20Israel%27s%20Water%20Reuse%20Approach.pdf.
25. TODDJ W. Contesting hidden waters: conflict resolution for groundwater aquifers[M]. New York: Routledge, 2014.
26. ROBERTS K. Fluid futures: integrating cultural values into water policy and governance[EB/OL]. (2024-11-14)[2025-01-10]. https://jointactionforwater.org/fluid-futures-integrating-cultural-values-into-water-policy-and-governance/.
27. KANWAL U. The role of Indus Water Treaty in fostering conflict and cooperation amidst water insecurities[EB/OL]. (2024-08-10)[2025-01-10]. https://strafasia.com/the-role-of-indus-water-treaty-in-fostering-conflict-and-cooperation-amidst-water-insecurities/.
28. KING M D. Water and conflict in the Middle East[M/OL]. Oxford:Oxford University Press, 2020[2025-02-10]. https://doi.org/10.1093/oso/9780197552636.001.0001.
29. KLAVER I J. Introduction: water and cultural diversity[M]//JOHNSTON B, HIWASAKI L, KLAVER I, et al. Water, cultural diversity, and global environmental change. Dordrecht: Springer, 2011.

30. LIM S S. Environment atsingspring's desalination plant ground breaking ceremony [EB/OL]. [2025-02-12]. https://www. nas. gov. sg/archivesonline/speeches/record-details/4cb93c14-4327-11eb-85f4-005056a7c31c.

31. LIPCHIN C, PALLANT E, SARANGA D, et al. Integrated water resources management and security in the Middle East[M]. Netherlands: Springer, 2007.

32. MARIN P, TAL S, YERES J, et al. Water management in Israel: key innovations and lessons learned for water-scarce countries[R/OL]. [2025-01-10]. https://documents. worldbank. org/en/publication/documents-reports/documentdetail/657531504204943236/water-management-in-israel-key-innovations-and-lessons-learned-for-water-scarce-countries.

33. MCCAFFREY S. The law of international watercourses: non-navigational uses Oxford monographs in international law [M]. Oxford: Oxford University Press, 2003.

34. MEKONG RIVER COMMISSION. The MRC Council study and the impacts of hydropower development on the Mekong[R/OL]. [2024-12-10]. https://www. internationalrivers. org/wp-content/uploads/sites/86/2021/03/tragic-trade-offs-the-mrc-council-study-and-the-impacts-of-hydropower-development-on-the-mekong. pdf.

35. MIRUMACHI N. Informal water diplomacy and power: a case of seeking water security in the Mekong River Basin [J]. Environmental Science & Policy, 2020, 114: 86-95.

36. COOLEY H, GLEICK P, WILKINSON R. Agricultural water conservation and efficiency potential in California—issue brief [R/OL]. [2024-10-26]. https://pacinst. org/publication/agricultural-water-conservation-and-efficiency-potential-in-california-issue-brief/.

37. NETAFIM. Sustainability review 2018-2019[R/OL]. [2024-10-30]. https://www. netafim. com. au/globalassets/sustainability/netafim-2018-2019-sustainability-review. pdf? v=4a63ed.

38. NETO S, CAMKIN J. Water diplomacy: more than a drop of wisdom[J]. WWP World Water Policy, 2024, 10(4): 1017-1022.

39. PACIFIC INSTITUTE. Water conflict[R/OL]. [2025-01-30]. https://www.worldwater.org/water-conflict/.

40. PAHL-WOSTL C, JEUKEN A. Integrated water resources management: theory and practice[M]. Dordrecht: Springer, 2018.

41. PELUSO N L, WATTS M. Violent environments[M]. New York: Cornell University Press, 2001.

42. PUBSINGAPORE'S NATIONAL WATER AGENCY. Our water story[EB/OL]. [2025-01-12]. https://www.pub.gov.sg/Public/WaterLoop/OurWaterStory.

43. REDCLIFT M R, GRASSO M. Handbook on climate change and human security[M]. Cheltenham: Edward Elgar Publishing, 2013.

44. SAXENA N, ISLAM M M, SHARMA D. Water pollution and remediation: a global concern[M/OL]. Cham: Springer, 2024[2025-02-01]. https://doi.org/10.1007/978-3-031-76301-4.

45. SCHMEIER S. Governing international watercourses: river basin organizations and the sustainable governance of internationally shared rivers and lakes[M]. New York: Routledge, 2013.

46. SCHMITT A, WERNERS S E. Flood risk management in Europe: innovation in policy and practice[M]. Netherlands: Wiley, 2016.

47. ZEITOUN M, WARNER J. Hydro-hegemony: a framework for analysis of trans-boundary water conflicts[J]. Water Policy, 2006,8(5):435-460.

48. SIEGEL S M. Let there be water: Israel's solution for a water-starved world [M]. New York: Thomas Dunne Books, 2015.

49. SHABAN A. Water resources in the Middle East[M]// SHABAN A. Satellite monitoring of water resources in the Middle East. Cham: Springer International Publishing, 2022:15-30.

50. HUNTJENS P, YASUDA Y, SWAIN A, et al. The multi-track water diplomacy framework: a legal and political economy analysis for advancing cooperation over shared waters[R/OL]. [2024-12-20]. https://siwi.org/wp-content/uploads/2018/01/THIGJ_The-Multi-track-Water-Diplomacy-Framework_Webversion-1-1.pdf.

51. SOKHEM P, SUNADA K, OISHI S. Managing transboundary rivers: the case of the Mekong River Basin[J]. Water International, 2007,32(4): 503-523.

52. STERN C V, SHEIKH P A. Management of the Colorado River: water allocations, drought, and the federal role[R/OL]. [2024-12-20]. https://www.congress.gov/crs_external_products/R/PDF/R45546/R45546.41.pdf.

53. TENNE A. Sea water desalination in Israel: planning, coping with difficulties, and economic aspects of long-term risks[R]. Israel Water Authority, State of Israel Desalination Division, 2010.

54. THORNE C R, Evans E P, PENNING-ROWSELL E C. Future Flooding and Coastal Erosion Risks[M]. London: Thomas Telford Publishing, 2007.

55. TSEGAI D, AUGENSTEIN P, HUANG Z, et al. Global drought snapshot 2023: the need for proactive action [EB/OL]. [2024-11-20]. https://reliefweb.int/report/world/global-drought-snapshot-2023-need-proactive-action-enru.

56. HIGH-LEVEL PANEL ON WATER. Making every drop count: an agenda for water cooperation[R/OL]. [2024-11-20]. https://thedocs.worldbank.org/en/doc/623971522268005372-0090022018/original/17825HLPWOutcome.pdf.

57. UNITED NATIONS ECONOMIC COMMISSION FOR EUROPE. Frequently asked questions on the 1992 water convention with the road map to facilitate accession processes [R/OL]. [2024-11-30]. https://doi.org/10.18356/9789210051262.

58. UNECE. Policy guidance note on the benefits of transboundary water cooperation : identification, assessment and communication [EB/OL]. [2024-12-20]. https://digitallibrary.un.org/record/810004? v=pdf.

59. UNITED NATIONS DEVELOPMENT PROGRAMME. Capacity development: a UNDP primer[R]. New York: UNDP, 2009.

60. ESCWA, ACSAD, FAO. Climate-proof watershed management design and resilience synthesis report[R/OL]. [2024-12-20]. https://www.unescwa.org/sites/default/files/pubs/pdf/climate-proof-watershed-management-design-resilience-english.pdf.

61. UNITED NATIONS OFFICE FOR DISASTER RISK REDUCTION. 2015 global assessment report on disaster risk reduction [R/OL]. [2024-12-20]. https://digitallibrary.un.org/record/4037965? v=pdf.

62. UNITED NATIONS. The United Nations World Water Development Report 2024: water for prosperity and peace[R/OL]. [2025-02-15]. https://unesdoc.unesco.org/ark:/48223/pf0000388948.

63. WANG H, HUANG G. Innovative strategies for flood control: lessons learned from global practices[M]. Florida: CRC Press, 2019.

64. WHEELER S A, NAUGES C, GRAFTON R Q. Water pricing, costs and markets[R]. Paris: Global Commission on the Economics of Water, 2023.

65. WOLF A T. The transboundary freshwater dispute database project[J]. Water International, 1999,24(2):160-163.

66. WOLF A T, STAHL K, MACOMBER M F. Conflict and cooperation within international river basins: the importance of institutional capacity[J]. Journal of the American Water Resources Association, 2003:1109-1026.

67. YOFFE S, FISKE G, GIORDANO M, et al. Geography of international water conflict and cooperation: data sets and applications[J]. Water Resources Research, 2004,40(5): W05S04.

68. WORLD BANK. Operation and maintenance strategies for hydropower: six case studies[R/OL]. [2024-12-20]. https://documents1.worldbank.org/curated/en/573221581011972528/pdf/Six-Case-Studies.pdf.

69. WORLD HEALTH ORGANIZATION. Guidelines for drinking-water quality: fourth edition incorporating the first and second addenda[R/OL]. (2022-03-21)[2024-12-20]. https://www.who.int/publications/i/item/9789240045064.

70. WORLD METEOROLOGICAL ORGANIZATION. Guidelines on flood management [R]. Geneva: WMO,2013.

71. AIVAZIDOU E. Water management in agriculture and industry: challenges, trends, and opportunities[J]. Sustainability MDPI, 2022,14:6.

72. XU X H, ZHOU Y J, CHEN X H. Project design management for a large hydropower station[M]// SAMADI-BOROUJENI H. Hydropower—practice and application. London: IntechOpen, 2020:1-20.

73. SIMONOVIĆ S P. Floods in a changing climate: risk management[M]. Cambridge: Cambridge University Press,2018.

74. ZEKRI S. Water policies in MENA countries[M]. Switzerland: Springer,

2020.
75. 胡文俊,孙岩. 世界主要国家的节水措施及其启发与思考[J]. 水利经济,2024,42(2):1-6.
76. LOWDER S K, SÁNCHEZ M V, BERTINI R. Which farms feed the world and has farmland become more concentrated? [J]. World Development, 2021,142(1).
77. UNESCO, WORLD WATER ASSESSMENT PROGRAMME. Water: a shared responsibility; the United Nations world water development report 2 [M]. New York: Berghan Books, 2006.
78. FAO, IFAD, UNICEF, et al. The state of food security and nutrition in the world 2017: building resilience for peace and food security [R]. Rome: FAO, 2017.
79. UNEP. Annual report 2022 [R]. Nairobi: United Nations Environment Programme, 2023.